Count Me In!
K–5

To Eli Freeman and Nathan Joseph, with love.
May you have teachers as dedicated and knowledgable as those who
contributed to this book.

Count Me In!

K–5

Including Learners With Special
Needs in Mathematics Classrooms

Judy Storeygard

Foreword by Karen Karp

Skyhorse Publishing

Skyhorse Publishing books may be purchased in bulk at special discounts
for sales promotion, corporate gifts, fund-raising, or educational purposes.
Special editions can also be created to specifications. For details, contact the
Special Sales Department, Skyhorse Publishing, 307 West 36th Street, 11th
Floor, New York, NY 10018 or info@skyhorsepublishing.com.

Skyhorse® and Skyhorse Publishing® are registered trademarks of Skyhorse
Publishing, Inc.®, a Delaware corporation.

Visit our website at www.skyhorsepublishing.com.

10 9 8 7 6 5 4 3 2 1

Library of Congress Cataloging-in-Publication Data is available on file.

Cover design by Scott Van Atta

Print ISBN: 978-1-62914-562-4
Ebook ISBN: 978-1-62914-911-0

Printed in the United States of America

Contents

Foreword

Teachers are the lever in making change. We know that teachers can increase student progress by taking on the study of the complex, multifaceted task of teaching mathematics. This can be accomplished, in part, through learning about what research suggests has worked to increase student performance. Yet teachers who seek effective practices in teaching mathematics to students with special needs will find that the amount of research-based strategies for mathematics instruction pales in comparison to the expansive array of strategies for use with students with special needs in literacy. Teaching mathematics is intellectually challenging, as is teaching students with learning disabilities. The combination can be doubly difficult. These are the very reasons why this book is so significant and its value so unique.

Simply put, Judy Storeygard focuses on the learning issue rather than the label—or formal diagnosis. By emphasizing the characteristics of learning needs, she focuses on an approach that helpfully responds to the needs of students across disabilities. Building on multitiered prevention systems used in many states, she moves away from consistently starting with a focus on the "gaps" in students' knowledge. She instead moves to identifying strengths in students' prior knowledge—a sounder first step. She emphasizes the behaviors students present every day in the classroom and finds ways to combat the tendency to step in and do the work for the students. In its place is this press for acknowledging and identifying capabilities and potential contributions from which to build. This welcome approach is the underlying framework for the book—in essence minimizing what the student cannot do and maximizing the student's opportunities to respond.

In *Count Me In!*, Judy steps out in front and illustrates how this change to more effective mathematics teaching can occur through the shared examples drawn from the work of classroom teachers. In the spirit of her last book, *My Kids Can,* she continues to hold high expectations for all learners by crafting chapters that move beyond the labeling of students. She details ways to uncover some of the complex barriers to learning challenges, and shows how teachers can respond to a variety of misconceptions on their own.

Count Me In! showcases authentic environments in classrooms using a feature called "Voices From the Field." Here, classroom teachers directly communicate their wisdom about the very purposeful choices they make. These conversations result in suggestions ranging from practical ideas that are applicable to Monday morning, as well as thoughtful, broad-based reflections that provide the foundations for an overarching philosophy about working with students with disabilities. The book addresses the challenges teachers confront every day in building the following for students:

- Capacity for learning mathematics
- Cognitive flexibility
- Ability to organize, plan, and self-monitor
- Development of useful strategies
- Skill in expressing mathematical ideas

By asking questions along the way, Judy reveals how to identify the organizational, behavioral, and cognitive skills necessary for students with special needs to derive meaning from a variety of activities. She then describes specific ways to provide additional support if any weaknesses in these skills are diagnosed.

Because approximately 80 percent of students' time in mathematics class is spent working on mathematics problems, Judy's emphasis on a problem-solving approach to instruction that promotes and supports sense-making is critical. Peppered with the verbatim dialogue from problem-solving sessions and samples of student work, she helps readers to see the moves necessary to help kids make sense of problem situations and to communicate their mathematical ideas. She paints vivid pictures of classrooms where the teacher's language and organization are shared in think-alouds; these bring the process of building a safe environment for learning and cultivating a culture of acceptance to the forefront.

It is important for all teachers to know, when reading this book, that it is not about the fear of facing the challenge of learning to teach mathematics to students with disabilities in a potentially new way. It is about the complex task of doing so—and how others who've gone before can support their colleagues. Teachers' voices are central to this work, and the mantra in one classroom, *effective effort plus time equals success*, holds true for teachers as well as students.

As one of the teachers in the book said to her students, "If you'd like help getting started, please come join us."

Karen Karp
Professor of Mathematics Education
at the University of Louisville

Acknowledgments

This book is a product of collaboration with a group of dedicated teachers, colleagues, and parents. Their insight and commitment has been essential to the process of conceiving, writing, and revising *Count Me In!*

In many ways, parents have inspired this book. Listening to their concerns about their children's education, and seeing their determination for their children's teachers to see their children's strengths as well as their needs, has greatly increased my awareness and understanding. I give special thanks to Nancy Isaacs, Laurie Brennan, Lenworth Hall, Janet Altobello, and Karen Economopoulos for their poignant and honest accounts of their children's educational experiences.

The Educational Research Collaborative at TERC provided me with funding to pursue publication. My colleagues, Arusha Hollister and Myriam Steinback, have been unfailingly generous in their support and astute feedback. Andee Rubin has been particularly helpful in advising me about the technology-related pieces, providing examples from her research project, INK-12: Teaching and Learning Using Interactive Ink Inscriptions in K-12; Curtis Killian has been an invaluable resource in creating graphics and offering general technological support. Karen Mutch-Jones and Amy Brodesky (Education Development Center colleague) have lent their wisdom about students with special needs.

I am also very grateful for my knowledgeable colleagues from the Professional Development Study Group who provided me with excellent comments on several chapters. In addition, Deborah Schifter, Marion Reynolds, and Connie Henry subsequently read revised versions.

In addition to the talented contributing teacher-authors, many other excellent teachers have played a part in informing my thinking and in allowing me to observe and write about their outstanding teaching. Excerpts from these observations are included in the book.

I want to thank Eileen Backus, Sara Gardner, Linda Jackson, Amy Monkiewicz, Lisa Nierenberg, Karin Olson-Shannon, Dee Watson, and Sara Wolff.

Jessica Allan and Cassandra Seibel at Corwin have guided and encouraged me throughout. Finally, I would like to thank my family for their love and support.

Publisher's Acknowledgments

We gratefully acknowledge the contributions of the following reviewers:

Roxie Ahlbrecht
Teacher Leader, NBCT
Second Grade, Math
Robert Frost Elementary School
Sioux Falls, SD

Carol Amos
Teacher, Math
Twinfield Union School
Kennett Square, VT

Sue Delay
Learning Support Teacher
Cedar Hills Elementary School
Oak Creek, WI

Susan German
Teacher, Math and Science
Hallsville Middle School
Hallsville, MO

Jennifer Harper
Fourth-Grade Teacher
Cavendish Town Elementary
Proctorsville, VT

Debra Howell
Teacher, Monte Cristo
Elementary
Granite Falls, WA

Diane K. Masarik
Assistant Professor of Teacher
Education, Math and Science
University of Wisconsin—Eau
Claire
Eau Claire, WI

Edward Nolan
Mathematics Department
Chairperson
Montgomery County PS
Market, MD

Renee Peoples
Teacher Leader, Math
Swain West Elementary
Bryson City, NC

About the Author

Judy Storeygard has been a senior research associate at TERC for the past 20 years, an independent, not for profit research and development organization whose mission is to improve mathematics and science education.

She has a long-term interest in and passion about the mathematics education of students with special needs and from underrepresented populations. Collaborating with teachers has been a prime focus of her work. She has also taught students with learning disabilities and behavioral disorders, and supervised graduate students in a moderate special needs certification program. As a member of the board of the Massachusetts Tourette Syndrome Association for over 15 years, she cochaired the Tourette Syndrome Association Conference for Educators from 1998 to 2002. At TERC, she has been a member of the ERC Fellowship Committee, an initiative that seeks recent PhDs or EdDs whose research focuses on enhancing teaching and learning opportunities in math and science for children, youth, and adults from historically nondominant communities.

Count Me In! is a continuation of the work she has done for the past decade. The Annenberg Challenge Fund and the National Science Foundation, Research in Disabilities Education division, have funded much of her work. Other roles that have resulted from these projects include contributing author to *Models of Intervention in Mathematics: Reweaving the Tapestry* (Fosnot, 2010); editor of the collection of video and written cases written with elementary mathematics teachers, *My Kids Can: Making Math Accessible to All Learners, K–5* (2009); contributing editor to *Working with the Range of Learners: Classroom Cases* in the *Investigations in Number, Data, and Space* (TERC) second edition; and coauthor for publications in the journals *Teaching Children Mathematics* (2007) and *Teaching Exceptional Children Plus* (2005).

About the Contributors

Zachary Champagne taught fourth and fifth grade in a large urban district in North Florida for 13 years. Currently, he is working on a two-year grant as a district facilitator for the Florida Center for Research In Science, Technology, Engineering, and Mathematics (FCR-STEM). He was the 2010 Teacher of the Year for his county and finalist for the Macy's Florida Teacher of the Year. He also was the 2006 recipient of the Presidential Award for Excellence in Mathematics and Science Teaching. He believes strongly in making mathematics concepts concrete and applicable to all of his students, and he works to show his students that "mathematics makes sense."

Nikki Faria-Mitchell is a third-grade teacher in the Boston Public Schools. She is also part of a project called Using Routines as an Instructional Tool for Developing Students' Conceptions of Proof—a collaboration of TERC, Education Development Center (EDC), and Mt. Holyoke College. She has also participated in and facilitated many Developing Mathematical Ideas workshops. She is especially interested in facilitating mathematical conversations that include all learners and allow her students to take ownership of their math learning.

Abbie Fox is a third-grade teacher in Natick, Massachusetts. She is passionate about helping kids apply their math skills in science and engineering and loves to incorporate technology projects into her teaching. She also facilitates lesson study with her colleagues, a form of professional development in which teachers codevelop lessons, observe one another's teaching, and take a close look at student's mathematical understandings.

Tiffany Young Frank is a preK teacher in the Boston Public Schools. She remembers being a young struggling math student. As a teacher, she has been committed to teaching all students, especially children of color, the intricacies of early math concepts with a focus on hands on instruction. In 2007, she was a Fund for Teachers fellow, and traveled to Melbourne, Australia (home to the Early Numeracy Project) to investigate effective research-based math instruction. She was also a part of rewriting and testing new math ideas for the TERC Investigations program for kindergarten students. She hopes her learning will enrich her teaching, her students, and her community.

Elizabeth Henshaw Harrington is a second-grade teacher on the north shore of Boston. As a person who struggled with math as a student, she was surprised to find she loves to teach math. She hopes her students will always love math and feel confident about the amazing work of which they are capable.

Sherri Neasman is a fifth-grade teacher in Boston Public Schools. She enjoys teaching all subjects; however, she most enjoys teaching math. As a child, she remembers "doing" math, but not being able to articulate exactly what was being "done." Because of this, she participated in the Title IIB math project, which sought to bridge the gap in the language of math learning between upper elementary and middle school grades. She is a member of the Grade 5 Assessment Preview Team whose goal is to examine the language of the exams and tests. She also tutors elementary and middle school math after school.

Lisa Nguyen is a fifth-grade teacher in the Boston Public Schools. She has also been a leader in several professional development projects related to elementary school math. She is dedicated to her practice in developing student mathematical thinking. Her focus is on developing students' ability to explain and challenge one another's mathematical thoughts by using math language.

Danielle Silverman is currently an elementary math resource room teacher for the Boston Public Schools. As a former math coach for the district, she continues to work collaboratively with the elementary math department to facilitate districtwide professional development on response to intervention (RTI) and the Investigations curriculum. She also facilitates Developing Mathematical Ideas workshops for Boston Public Schools and works as a consultant for TERC. In addition to advocating for special education students and their mathematical learning needs, she enjoys spending time with her new baby boy, traveling, reading, photography, and spending time on Nauset Outer Beach on Cape Cod.

Heather Straughter taught fourth and fifth grade in the Boston Public Schools for eight years. She looped with her students and thoroughly enjoyed teaching them for two years. She was also a Developing Mathematical Ideas facilitator and leader in professional development for other teachers. She left teaching in 2005 to become a stay-at-home mom and is enjoying using what she knows about teaching with her now elementary-aged son.

Ana Vaisenstein has been working in the Boston Public Schools since 1998. She has taught kindergarten through fifth grade to English language learners and has coached teachers in mathematics. She is currently teaching a third-grade sheltered English immersion class. She enjoys observing students explain their ideas to one another and make sense of problems together.

Introduction

Never mind the labels; my child has learning difficulties, what can we do about them?

This statement by a parent whose child with special needs has been diagnosed over the years with attention deficit hyperactivity disorder (ADHD), Asperger's syndrome (now included with autism spectrum disorders), nonverbal learning disorder (NLD), and pervasive developmental disorder (PDD)[1] reflects the purpose of this book. The book's intent is to provide a resource for teachers by addressing some of the common learning characteristics ascribed to students with special needs. The information and episodes in the book go beyond the labels and uncover some of the complexities, strategies, and reflections from teachers about their mathematical work with these students.

Stories from parents of students with special needs, such as the parent quoted at the beginning of this section, have inspired this book. These parents are often frustrated that their children's education focuses on deficits that accompany labels and ignore their child's strengths. If a child has been labeled as having ADHD or PDD, the characteristics of that syndrome tend to dominate Individualized Education Program (IEP) meetings and discussions. The same parent quoted here dreaded parent conferences because they focused almost exclusively on her child's problematic social skills and disruptive behaviors. She felt the implication was, "Why haven't you (the parents) fixed this problem yet? We (the teachers) keep telling you what's wrong, and we're frustrated that it doesn't change. Unless you can fix your child's behavior issues, we're not going to be able to do much to teach him." The teachers wanted no ownership of any behavioral problem, and saw no connection between these issues and how they were teaching. Although the child has many academic strengths, including mathematics, the conversation did not include ways to work together to make use of these strengths in a way that might begin to build the child's social skills.

Conversations with parents illustrate the deficit mentality that their children have faced in schools. A parent of a child diagnosed with ADHD revealed that the majority of the time in parent conferences was devoted to discussing the child's behavior, not his academic strengths. She and her husband had to bring their child's strengths into the conversation. The school described him as a "borderline" student, yet at home the child was reading fluently, could solve math computation problems, and used logical reasoning. He also played a musical instrument and built model cars. In his afterschool program, few behavior problems arose. The environment of his third-grade classroom, however, was not conducive to his learning. His classroom had over 30 students with most instruction in large groups that relied heavily on auditory transmission. The students had only 15 minutes for lunch and no recess. The parents knew that their son learns best using representations and models, but these were not available to him; they also knew that he has trouble continuing to pay attention without more time to move around. Instead of considering how the academic environment might be modified to build on his strengths, the school has labeled this third-grade child as a problem because of his impulsivity, and even suspended him. The child is now anxious about school and has frequent stomachaches. The parents are sad that the school doesn't see their "lively, curious, kind child who would give you the shirt off his back."

One parent of a child with an autism spectrum disorder (ASD) reported that her child has many math-related skills. For example, he has an amazing memory for directions; he can find his way anywhere. He completes complex puzzles easily, and can follow directions to build Lego models. Yet in the math class in the special school he attends, expectations for math learning have been low. He has never been introduced to word problems and was only taught to count by ones until his mother taught him to group by 5s and 10s and use a number line. Despite his visual-spatial skills, his instruction in geometry has been limited to shape and angle recognition. "How can his skills be used to build math knowledge?" his mother wants to know. Unfortunately, this question has not been addressed during his schooling. Mathematics instruction that focuses on sense-making too often takes a backseat in special education settings. Instead, instruction is often confined to practice on rote skills and on life skills such as time and money.

Making Sense of Mathematics: The Guiding Principle of This Book

The stories from these parents describe environments in which their children's mathematical abilities are not being fully developed. In contrast, the teachers who have contributed to this book all expect their students to learn mathematics with understanding and are searching for strategies that can help their students with special needs make sense of mathematics. They embrace the standards for mathematical practice for all learners laid out in the Common Core State Standards for Mathematics (www.corestandards.org) that focus on making sense and developing mathematical reasoning. In order to develop mathematical skills and concepts, children need to make sense of mathematics, and teachers need to observe and assess carefully so they know how children learn, what strengths they have, and where they need support. The diagnostic labels their students with special needs receive often do not help teachers in their daily classroom preparation. They need instead to think about the particular learning needs students exhibit every day, and develop strategies so that their students with special needs can learn mathematics.

Focus on Labels: Ableism

Not only are many diagnostic labels of limited use, but a focus on labels also means that students are often seen only through the lens of disability instead of as individuals who have strengths, but may need support and opportunities to learn. The teachers who are contributors to this book have the goal of helping all of their students, including those who are labeled with disabilities, access mathematical ideas in ways that meet their individual learning needs.

Focusing on labels leads to ableism, a prejudice against people with disabilities, both in how they are seen and educated. Ableism is a perspective that asserts that "it is better for a child to walk than roll, speak than sign, read print than read Braille, spell independently than use a spell-check, and hang out with nondisabled kids as opposed to other disabled kids" (Hehir, 2002, p. 3). While Hehir and his colleagues are proponents of offering these students the most up-to-date medical and technological assistance, they also assert that ableism leads to a school culture in which teachers are expected to "fix" students' disabilities instead of minimizing the impact of the disability while maximizing the child's opportunities to participate (Hehir, 2007). Viewing a child through a label often takes precedence over *maximizing his or her opportunities to participate* in engaging in and learning important mathematical ideas.

Inconsistency of Labels

Not only can labels limit students' possibilities, but there is also a great deal of evidence that guidelines for labels are unclear; thus, they are applied inconsistently. The research on disproportionality supplies evidence of how labels are applied unevenly.

Many studies of disproportionality have found a lack of clear criteria for referrals. African American students are more likely than white students to be placed in emotional disturbance (ED) and mental retardation categories, and less likely to be placed in LD categories (Skiba, Poloni-Staudinger, Gallini, Simmons, & Feggins-Azziz, 2006). Hispanic students are underrepresented in the ADHD category (Schnoes, Reid, Wagner, & Marder, 2006).

While the Individuals With Disabilities Education act (IDEA) calls for students to be in the "least restrictive environment" (LRE) (Lipsky & Garner, 1999), the disproportionate representation of minority students and English language learners in special education and in segregated settings has been well documented (Artiles, 2003; Losen & Orfield, 2002; Valenzuela, Copeland, Qi, & Park, 2006). Further, the overrepresentation promotes racial stereotypes; little attention is paid to heterogeneity between and within culturally and linguistically diverse students and disability groups (Waitoller, Artiles, & Cheney, 2009). These inconsistent uses of labels show how little use they are to educators whose goal is to plan instruction that meets their students' learning needs.

Labeling Impedes Differentiation

Negotiating the territory of supporting students with special needs has become even more fraught with pressures in recent years. Teachers, already under stress, often must sort out a set of competing mandates. They are told that students have to perform well on standardized tests, yet they have to "differentiate" instruction and take major responsibility in response to intervention (RTI). RTI is a tiered system designed to improve learning through instruction, assessment, and intervention (National Center on Response to Intervention, 2010). While initially used in reading instruction, RTI is becoming more common in mathematics classrooms. Available personnel and resources to implement this initiative vary widely among school districts.

Ironically, while teachers are increasingly asked to differentiate instruction, students with special needs are often described as having a similar set of characteristics and are defined by their diagnostic label, thus complicating the task of differentiating. Rather than viewing learners' starting places as "gaps," teachers need to understand learners' prior experiences and knowledge (Wilson & Peterson, 2006). Instead of just receiving a score on a standardized test for their students with special needs, teachers need assessments that inform instruction, to help them understand how students learn and identify areas of strength as well as weaknesses.

Including a Variety of Instructional Strategies: The Approach of This Book

The teachers who contribute to this book do not teach in a one-size-fits-all way.

As Fosnot (2010) states, "We cannot get all learners to the same landmarks at the same time, in the same way, any more than we can get all toddlers to walk at the same time, in the same way" (p. 23). Students come to mathematics class with different skills and needs, and it is the teacher's responsibility to meet them where they are. Teachers who meet this responsibility use formative assessment to identify students' strengths and areas of support, designing and adapting their instructional and assessment approaches accordingly and offering a variety of models and representations that will help students access the mathematics. Their teaching practices include a repertoire of instructional strategies, and they make sound decisions about how and when to use them. Their mathematical goals for their students are always to build mathematical understanding.

Some new resources are available to teachers. Teachers now have information to draw on, such as the universal design for learning (UDL) resources that support the development of flexible instructional goals, methods, materials, and assessments (Rose & Meyer, 2009). With the explosion of new technologies, teachers' classroom practices may include the use of these tools, whether interactive whiteboards or tablet computers, smartphones, or other devices that have easy access to video and Internet capability. Although these devices are still not universally available or able to always be supported in schools, they have the potential to provide flexibility to teachers in presenting content in multiple ways to meet the needs of a range of students with diverse needs. Using technology and other models and representations in this way supports students' strengths and increases their access to meaningful mathematics.

Organization of This Book

This book is intended to provide teachers with some examples and strategies that will help them meet the needs of their students with special needs.

While one book cannot cover all of the learning needs of students with special needs, we have chosen the most common challenges that teachers face in teaching mathematics in inclusive classrooms:

- Supporting students in expressing mathematical ideas
- Helping students to build the capacity to attend to and focus on mathematical ideas
- Developing students' cognitive flexibility
- Developing strategies for students with memory difficulties
- Building students' abilities to plan, organize, and self-monitor in mathematics class

Although specific diagnoses are often associated with these learning difficulties, we focus on the specific learning issue and how to foster mathematical learning instead of a diagnostic category. For example, instead of including a chapter that discusses autism spectrum disorders (ASD), we have chosen to examine the aspect of concrete thinking that often goes with these disorders, but can also be a feature of other disabilities, such as nonverbal learning disorder (NLD), post-traumatic stress disorder (PTSD), or learning disabilities in general. Thus, the chapter is about developing students' cognitive flexibility and focuses on strategies to foster students' understanding of concepts—strategies that can be helpful to students with many labels. Similarly, the chapter on developing students' abilities to attend and focus includes students who may need this support for a variety of reasons and with a variety of labeled diagnoses (e.g., ADHD, ASD). Definitions of specific conditions and terms are in the Glossary and Resources section.

Each chapter includes an introduction to the special need, some perspectives about the topic from research, and examples of strategies that teachers have used, including extensive teacher written episodes called "Voices From the Field."

What does it look like to teach mathematics that focuses on sense-making to a range of students in inclusive classrooms? In general, teachers have many more sources of support to consult for literacy instruction than for mathematics. By placing the specific learning behaviors that can impede progress in mathematical learning front and center, we hope to bring forth instructional and assessment strategies from the classrooms of practicing teachers along with their reflections about how to discover and emphasize students' strengths as well as address their learning challenges.

Note

1. See the Glossary and Resources section for definitions of these diagnostic and special education–related terms.

1

Building a Culture of Acceptance in the Inclusive Mathematics Classroom

The ability of students to learn mathematics with understanding relies on a classroom community in which all students are expected to learn through active participation, and teachers provide support to engage all students in mathematical tasks.

The work of Stigler and Hiebert (2009) in reporting on the Trends in International Mathematics and Science Study (TIMSS) video describes features of inclusive mathematics communities in which all students are learning to make sense of mathematics. The videos from Japanese classrooms show teachers who include confusion and frustration as part of the learning process, give students time to sort through what puzzles them, and offer support as needed. The Japanese teachers who were videoed welcome differences in the class because they often lead to a range of ideas that provides an entry point for discussion and reflection. All students are given the opportunity to learn the same mathematics, but teachers understand that different strategies will make sense to different students, and that not all students will learn the same things from each lesson.

Like the Japanese teachers in the TIMSS, the teachers who have contributed to this book create a culture of high expectations and acceptance of differences in their classrooms that facilitates the learning of mathematics. They make sure their students know that they are expected to take responsibility for their own learning and support one another as learners. Further, they expect their students with special needs to learn along with their peers. They create a culture based on respect and acceptance of differences in which students feel safe to take risks and to admit frustration and confusions. In fact, one teacher refers specifically to taking risks when she calls on children to discuss their strategies. When she calls on a child, she asks, "Do you want to take a risk or do you want to wait?" If someone makes a mistake, she might say, "Don't feel badly if you make a mistake. Many of you find this problem difficult." Finally, these teachers provide multiple points of entry based on their observations of what has worked for the different learners in their class.

Teaching for understanding in an inclusive mathematics classroom is not easy. Many teachers did not learn to make sense of mathematics when they were in school, and the process of learning to teach in this way is complex. Ms. Walker, an early childhood teacher in an urban school system, writes passionately about her own negative experience as a mathematical learner, how she was made to feel that she couldn't ask questions and couldn't learn mathematics. She describes how she is determined to create a supportive atmosphere and recognize the strengths of all of her students.

Voices From the Field

A Teacher Confronts Her Math Anxiety

I am a teacher in an inclusive classroom in a large public school system where over 80 percent of the children are students of color. Our classroom is a community of learners, and we try and develop the individual talents of each of our students. I have taken on the job as my personal responsibility to see the light in each one of my students no matter where they are from or what challenges are before them.

Mathematics is a subject that I've always liked, but as a student I remember that I was afraid to approach the teacher and ask for help. I didn't want anyone to know I didn't fully understand what was going on. In high school, our math classes were tracked and I remember being in "dumb" math, as we labeled it—the class with the students who needed more time to understand. In college I took a math course for teachers who were to teach math at the elementary level and did horribly. I asked the professor for help this time, but he made me feel so inadequate that I was not willing to ask for support again.

At the beginning of my teaching career, I was nervous about being in front of young children and teaching math. The pain of my personal story kept appearing. I was able to get them to memorize and regurgitate whatever I was telling them. This "worked" for a few years until I began taking courses for the math curriculum that I needed to teach. I found out that not only did I have to know the content, but, in addition, I needed to know how my students were learning math ideas. It was a learning experience both for me and my students. I learned from math coaches as I watched them teach lessons and processed the lessons with them. They helped me navigate the curriculum for my students, and they guided me to understand the math for myself. I built up my confidence and have spent a great deal of time analyzing my work. The teaching of math has become so important to me that I try to be involved with as many opportunities as I can to share ideas. I continue to learn about how young children understand math in order to help them build a mathematical foundation based on developmentally appropriate concepts. In contrast to my own experience of isolation and anxiety in math class, I create a mathematics community in my classroom that is based on testing out ideas, asking questions, feeling safe to make mistakes and learning form each other.

Through her own experience as a teacher, Ms. Walker came to learn the pleasure and importance of making sense of mathematics and has worked hard to create a culture of sense-making in her own class that is very different from her own experience learning math.

Recognizing Differences and Supporting Strengths

As indicated in the examples from Japanese classrooms, recognizing differences among students can be an opportunity to surface both a range of solutions and confusions. To celebrate differences, they must be acknowledged so that both the teacher and the students can support and learn from each other. The following classroom examples illustrate teachers who make a point of including everyone in ways that both recognize differences and use student strengths as avenues for learning.

When she was a new teacher, Ms. Gordon hadn't yet learned to acknowledge student differences in a way that supported learning. Here she describes the beginning of her journey and reflects on the consequences of not being explicit about learning differences and the supports that different students sometimes need.

Voices From the Field

"It's Because We're Bad at Math."

My first attempt at differentiating math instruction was a disaster. I was a teaching intern, and our third graders had been working on addition. A little more than half of them were confidently adding two-digit numbers, and the rest needed help developing strategies.

The solution was simple. After our mini-lesson, I'd split the class in two. I would work with the kids who already had solid strategies, helping them to show and extend their thinking. The rest of the class would go with my mentor teacher, Anne, to the adjoining room. Together, they would work more on understanding place value and breaking up numbers by 10s and 1s.

I stood in front of the class and explained the plan. "So, Thomas, Eden, Olivia, Mark, and Sarah, please take your folders and follow Anne." At that moment, Sarah's body stiffened and her eyes filled with tears. She looked up at me with an angry, hurt expression. "Why do they get to stay?" she asked. "Why do we have to go in the other room? It's because we're bad at math. It's because we're stupid."

Every student turned to look at me, and I had no idea of how to respond to Sarah's accusation. I fumbled, trying to explain. "No, of course not. People learn in different ways . . . we can do a better job of teaching you . . . small groups will help . . ." I had unwittingly realized one of a teacher's greatest fears. I had made a student who was having difficulty feel like a failure. I had confirmed her suspicion that one group of students was "good at math" while another was "bad at math" as clearly as my teachers had when they put struggling learners in the back of the class when I was young.

There were several problems with my approach that day. First, I publicly announced who would be in what group without explanation of the purpose. Second, by sending one group to another room—which I had wanted only to minimize distractions—I had physically segregated my class for the remainder of the math period. I had presented no opportunities to move from one group to another. Third, and most important, I had sprung this new structure on our class with no discussion about learning differences, no expectation that different students sometimes needed different amounts of help and support.

Sarah had drawn a logical conclusion from my actions: I didn't believe she was smart enough to work alongside the rest of the class. I had confirmed her long-held fear that she was simply "bad at math."

As we will see in the next section, Ms. Gordon learns an important lesson from this incident that leads her to think carefully about how she wants to create a supportive classroom community.

Ms. Thompson, a teacher who has high expectations for all of her students, expresses her philosophy as follows: "I believe in making public the things that are difficult for all of us, myself included, as a way of teaching empathy and support in the classroom." She organizes her math class with set routines that work for everyone, including Michael. Michael, a student in her fourth-grade class, responds well to a structured environment. He needs and wants clear and consistent routines, and becomes anxious if he is uncertain about what to do either academically or socially. Ms. Thompson and Michael thus developed strategies to diffuse his anxiety, ones that were clear to him and to the rest of the class. She included his classmates in openly acknowledging his differences and enlisting their cooperation in helping Michael feel comfortable, while also making them aware of his strengths.

Voices From the Field

Lessening Anxiety and Supporting Learning

In order for Michael to make sense of mathematics and, ultimately, become a successful mathematics student, I had to create a learning environment where he could feel safe and comfortable in all aspects of the day. I set up daily routines for the class. Typically, we started math with the students sitting on the rug, each with a white board, a pen, and a paper towel. I posed problems for students to solve on the whiteboard.

We developed strategies to help Michael when he became anxious or frustrated; for example, we would suggest he take a walk in the hall and not reenter the room until he was "ready to work." When he was having trouble working with a partner or when he had made a small computational error, I would question him, as his third-grade teacher had, "Is it really that serious?" Once these strategies became routine, Michael became less apprehensive in math class and was able to work more productively.

As he progressed through his fourth-grade year, I came to more fully appreciate Michael and his learning style. I recognized what worked for him and, more important, Michael began to grasp the expectations of the classroom and understand that they also applied to him. I believe that this was a key component for Michael finding success. In earlier grades, he had grown accustomed to lower expectations than the rest of the

(Continued)

(Continued)

class, and knowing this added to his feeling of "I can't do it." By raising the expectations and slowly teaching him the skills to meet these expectations, I started to see that Michael was making progress, and even more important, he began to see his own progress.

Fully accepting Michael by acknowledging his strengths and contributions as well as his needs is an important component of the inclusion model of our school. The teachers at Montgomery are explicit with our students that fairness means giving people what they need, not giving everyone the same thing. To put this into practice, I had conversations with the whole class. For instance, we discussed how we could help our friends if they get upset. Sometimes I had conversations with a smaller group of students about specific issues related to Michael and other students, such as, "How can we help Michael get out of his negative behaviors? If we feel we can't handle it, what do we do?" Michael's classmates learned about the ways that he became frustrated and they were able to support him. They were patient with him and patient with me when I needed to spend time with him, and several students were able to redirect him before he became overly anxious and acted out. Both the students and Michael could see that by lessening his anxiety, he was able to contribute strategies and ideas to our math conversations that benefited everyone.

Another teacher, Ms. Neal, makes a point of including all of her students in math conversations, recognizing situations in which they can make positive contributions with her support. For example, she calls on Sam, one of her students who struggles with cognitive flexibility, when she sees his hand raised to share a strategy. If he gets off track, she brings him back by asking, "How did you begin?" To bring other students into the conversation, she might ask, "Who began the same way?" When choosing someone to read a math problem out loud, she thinks about Kevin, a student who has some special needs, but who is an outstanding reader.

Both Ms. Thompson and Ms. Neal address the differences of their students by supporting them and finding ways that allow their successful participation in classroom mathematics.

Making the Classroom a Safe Place

Students with special needs often have not had positive experiences in school. They may be reluctant to contribute, fearful of making mistakes. As Ms. Walker reveals through her own experience, this anxiety is not conducive to learning math. Often a self-fulfilling prophecy results: the more the student does not contribute, the more likely expectations for the student's performance are lowered, which in turn, lowers the student's self esteem. These teachers spend a great deal of time in the beginning of the year making explicit the code of conduct in the class that will make everyone feel comfortable. When students find their classroom to be a safe place to make mistakes, to ask questions, and ask for help, then they are less likely to hide their confusions and struggles and to reach out and accept the assistance they need. Ms. Thompson defines her own role with the children to make it clear that she is there to help students learn:

> I want to know who is stuck at home tonight so I can help you Monday. That's my job. If you do your homework with someone and they do the work, I won't know if you're stuck, and I'll move on and you'll be more confused.

We return to Ms. Gordon as she continues her journey to establish a classroom community that acknowledges and supports differences so that students are comfortable expressing confusion and asking for help.

Voices From the Field

The No-Secrets Math Classroom Part 2

Today, extra help is an expectation—not an exception—in my classroom. After a whole-group mini-lesson, I subtly pull aside the students I know need help—sometimes, but not always, including special education students—and I let the class know what's going on. "At the round table, I'm going to do some more problems like this with a small group. If you'd like some help getting started, please come join us." Typically, between two and four additional students will join in. And it's not always the same students. Even high fliers—kids I wouldn't suspect of having difficulty—will come to the small group to clarify a question or to gain confidence in a new skill.

Students feel comfortable doing this in part because we've had conversations since the beginning of the year about learning differences.

Ms. Gordon: So far this year, we've talked about different school tools that kids need to be successful. Who would like to share about a tool they're using?

(Continued)

(Continued)

Samantha: I am using a pencil grip because it makes my handwriting neater. [Half a dozen students give Samantha the "me too" hand signal. This happens as each student shares.]

Josh: I use a slider to help me in reading so I don't lose my place.

Harry: I have a private office space so I can focus and not get distracted.

Madeleine: Some kids have checklists for checking over their work.

Ms. Gordon: That's right. We know that kids need different tools to be successful. Because we're all different people! People also need extra help sometimes when something is hard for them. Think for a moment, and put your hands on your knees if you can remember a time when you needed help. [I pause while students think. Just about everyone puts their hands on their knees.] Raise your hand if you'd like to share what you remember.

As students share, it becomes clear that everyone needs help sometimes (even teachers). Then I make my teaching point:

This year, I and the other teachers in our classroom will give extra help to kids during math. Does this mean that the kids I'm helping aren't good at math? Absolutely not. When we put in effective effort and time, we can achieve anything. You will all be challenged this year, sometimes in math, sometimes in reading or writing or making good choices, but you will all be successful.

The message, *effective effort plus time equals success,* is a constant theme in our classroom. So when I ask students to join us if they need extra help, they do, and that extra help carries no stigma, no feeling of being "bad at math." We have no secrets in our math classroom. Because we talk about it, everyone knows that new ideas come quickly to some and more slowly to others. Everyone knows that different people have different strengths. And everyone knows that we all need help sometimes, and that it feels good to get the help you need.

Ms. Gordon's ability to reflect on her own practice leads her to create a learning community in her classroom in which students feel comfortable acknowledging areas for improvement and asking for help in order to work toward their learning goals. By making everyone's needs public, she creates an entry point for discussion and for support. She shifts the responsibility for learning to the students instead of preassigning groups.

Promoting Responsibility

An emphasis on responsibility that teachers like Ms. Gordon promote represents a sharp contrast to the "learned helplessness" that affects students with special needs when adults who work with them "tell them what to do." When adults simply tell them what to do, they see learning as outside of themselves and, as a result, are not familiar with grappling with problems and trying to make sense of the mathematics. Students who take responsibility for their own learning are more likely to learn math with understanding because they are more likely to rely on their own knowledge instead of the knowledge of others. Teachers who promote responsibility promote both individual responsibility and responsibility for other students in a community of learners.

Even the way the classroom is physically organized can communicate to students the importance of shared responsibility. When students are organized in groups of four and desks are faced toward each other, it can facilitate their academic and casual conversation within all subject areas. Students can also develop a sense of ownership within the classroom through jobs and leadership roles to create an environment in which they are forced to take care of the classroom together. Some classroom jobs that promote student ownership of their own mathematics learning include team leaders whose job entails telling other students their tasks and managing student collaboration, before or after school, and homework checkers whose job entails checking homework every day. Further, some teachers assign students a day during which they take charge of the morning meeting when they discuss content objectives of the day.

Ms. Robinson lets her fifth graders know that they need to build a strong foundation in fifth-grade math for their middle school "house of math." She tells the class,

> Your job is to have a solid fifth-grade year. I don't know you're struggling if you don't tell me. Everyone is a math person. They "rock" in some areas, but they may struggle in some areas. Where do I struggle? You should be asking yourself that. My job is to make you rock.

When Ms. Robinson teaches, the students who need help gather on the rug. She works with her "rug rats," as she affectionately calls them, some at the beginning of the class, some come later if they're stuck. If she is circulating, students hold up three fingers if they need her, and one finger if they have something to share. When students help each other, she urges them to ask questions, and give examples, but not to tell the answer because that is not helping. Before a test, she writes two sentences on the board: *I believe in you. Do you believe in you?* After a unit, she asks students reflection questions: Was the unit easy? What made it easy? If it was hard, what would have made you understand it better? Ms. Robinson's approach makes the goals of math class clear to her students, and they understand both her role and their roles. Asking for extra help is a built-in routine of her classroom, and students take the initiative to seek support.

Acknowledging Frustration

Students who struggle with mathematics can become frustrated. Although a certain amount of frustration can promote learning, when frustration leads to various degrees of anger, this clearly impacts their ability to learn. It is important, then, to face these feelings of frustration and anger openly and to work on ways in which students can regulate their responses.

Ms. Thompson helps her students regulate their behavior with the use of an "anger thermometer" (see Figure 1.1). The anger thermometer is created as a whole class and is different for each class she teaches. It is important that the anger thermometer include ideas that are meaningful and organic to the particular students she is teaching.

Ms. Thompson creates a visual representation of a thermometer with different lines and categories matching particular items with appropriate responses. For instance, being rushed might lead to frustration while someone physically hurting a student might make that student furious.

Although the whole class creates and refers to this representation, the anger thermometer is particularly effective for those students who need the most help with self-regulation. This anger thermometer serves as a gauge for students to recognize if their reactions were appropriate or if they were in fact overreacting. Oftentimes, when students feel they are struggling or find themselves getting confused or unsure, they may overreact. They may get very upset and shut down. They may become furious when in fact it is something that is only worth being frustrated over. The thermometer may help students

Figure 1.1　Anger Thermometer

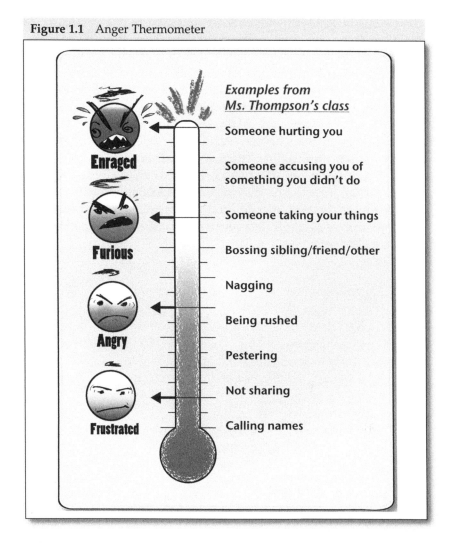

recognize that furious reactions are appropriate for when someone is hurting them or taking their things, not for needing extra help with a math problem. At times during math class, Ms. Thompson just points to the anger thermometer if a student has become frustrated, and the student is able to express a more appropriate response.

Summary

The chapters that follow include many of the principles and strategies introduced here. The teachers who have contributed to this book strive to establish a community of learners in their mathematics classrooms in which all students are expected to and can learn. As Ms. Gordon's class theme—*effective effort plus time equals success*—indicates, these teachers expect and work hard to engage everyone in making sense of mathematics. Rather than hide student differences, these teachers acknowledge them and structure their classrooms to provide support. Asking for help is not a stigma in these classrooms, it signifies that students are taking responsibility for their learning. The examples in this chapter reveal the purposeful choices these teachers make to build their community, whether it's Ms. Thompson's anger thermometer, Ms. Robinson's rug rats, or Ms. Gordon's round table. Teachers such as Ms. Walker and Ms. Gordon acknowledge the complexity of the task at hand and reflect on their own teaching; there are no easy solutions to addressing diverse student needs in an inclusive classroom.

2

Supporting Students in Expressing Mathematical Ideas

There is too much language in these NCTM Standards-based mathematics programs for students with special needs.

My students with special needs cannot explain their thinking.

Statements like these have often excluded students from an opportunity to engage in mathematical thinking. While expressive language problems present a challenge in mathematics learning as well as in the language arts, it is nevertheless important that these students have access to meaningful mathematics. As Ms. Williams, one of the contributing authors to this chapter, expresses, allowing students to verbalize their thinking in math class facilitates their ability to reason mathematically.

> Several years ago, a first-grade student of mine asked me why I ask so many questions. At the time, my response was that as students talk about their ideas, I am able to better understand how they make sense of those ideas and in turn it helps me with my teaching. I added that as they explained their ideas

to one another they become more aware of their own understanding. My students seemed satisfied with this explanation, and so did I. A few years later, however, I realized that there is more to math conversations. As the children and I talk about their strategies and thoughts, we clarify the use of our words and agree on a common vocabulary: we negotiate meaning. In doing so, all students learn to express ideas clearly. This is a process that unfolds over time. As children present their ideas, define and redefine them, the vocabulary and expressions that accompany them emerges and becomes part of the classroom culture. (See Ball & Bass, 2003.)

I find that focusing on math conversations helps all of my students, those who are comfortable expressing themselves, those who are learning a second language, and those who need support in expressing their thoughts.

Issues Related to Expressive Language Difficulties

There can be a variety of explanations for students' expressive language problems. For some, it is their primary area of difficulty. For others, it is a part of another condition. For example, 54 percent of children with ADHD may exhibit expressive language problems as opposed to 2 to 25 percent of a typically developing population. These children also have a higher rate of dysfluent and disorganized speech (DuPaul & Stoner, 2003).

Second language learners are a large category of students who may struggle with expressing their ideas mathematically. Too often English language learners are considered as a homogenous population when there is a great deal of variability in native language background, demographic contexts, and instructional opportunities. There is a dearth of knowledge about students learning English, especially those who are experiencing difficulty. The gaps are particularly apparent in these areas: (1) identification classification and assessment, (2) normative developmental trajectories of vocabulary and literacy skills, and (3) the context in which instruction takes place (Artiles & Klingner, 2006; Lesaux, 2006).

The following questions are important in evaluating the language of English language learners: What role does culture play? Is diversity of language expression valued and contextualized in terms of including the culture, history, values, experiences, and sociopolitical realities of the speakers? What might be the effect on the child's language if there is a cultural disconnect between teacher and child? When English learners are classified as having expressive language disabilities, how much of the difficulty stems from the process of acquiring a second language and to what extent is an expressive language learning disability involved? (Dyson & Smitherman, 2009).

Effective Strategies

As with the other chapters, the goal here is not to discuss the various labels and conditions, but for teachers, like Ms. Williams, to share strategies they have developed to support students to become comfortable with mathematical language and to communicate their thinking. They describe how they encourage students to use familiar language to express themselves and how they use this language to help solidify the students' knowledge of concepts before they introduce more formal terms. They help students make connections to mathematics by bringing in prior experiences that relate to their students' lives. They pay careful attention to how they ask questions or offer explanations, and if students do not seem to understand the information, they figure out where the confusion lies and offer alternative wording. In addition, they make use of manipulatives and representations to illustrate concepts and allow students to demonstrate their ideas. Finally, they provide structured support, such as sentence frames, to foster students' ability to express themselves mathematically. These strategies are effective for a wide variety of students who have difficulty expressing their mathematical thinking, no matter what the reasons.

Making Connections

When students struggle with expressive language, teachers find that making connections with their prior experiences gives them a "hook" to facilitate communication.

Students who have expressive language problems or are learning English often need a jumping-off point into a new concept or theme. In language arts, for her students who are English language learners, Ms. Robinson, uses read-alouds or stories to introduce concepts or a theme. She has had a more difficult time finding stories to use for math that connect to the lives of her diverse urban students, but she finds that using contexts from other subjects, such as social studies, works well.

To introduce division, Ms. Robinson refers to the social studies unit on slavery that the class is studying. She and the students have discussed the unit extensively, and her students expressed great interest in the topic. She wants her students to understand the horror of slavery. How many slaves were packed into the ships? She directs four students to squeeze onto a rug remnant to illustrate how tight the space was on the ship. She puts four remnants with sixteen students on them altogether, and the class goes on to figure out how many groups of sixteen would fit on a boat that held one thousand people. By using a situation they have studied and that was meaningful and then acting it out as a class, the appalling conditions of slavery become more vivid for the students, and division becomes something concrete.

When Ms. Robinson's class studies fractions, her students construct pizzas with different toppings out of construction paper. All are the same size, but the students choose different denominators and divide their pizzas into slices accordingly. In addition to the pizza activity, she uses familiar contexts so *all* students, not just the ones with language difficulties, understand to which *whole* the *half* refers. (e.g., *Half of our class of 24 and half of Ms. Connor's class of 32 is going to art, while the other half of each class goes to gym. How many students will be in art class, and how many will be in gym class? Or, Our class is dividing into teams for a soccer game, half will wear red, and half will wear blue. How many of each uniform will we need?*)

Tying mathematical language to familiar experiences is important in unpacking students' confusions about mathematical terms and how familiar words are used in mathematical contexts. The word *slope*, referring to a line on a graph, often does not make sense to some of Ms. Robinson's students who are English learners until she makes the analogy with a hill. The steeper the slope, the harder it is to walk up the hill. When *rate of change* comes up, Ms. Robinson returns to the science unit on plant growth. She asks the class, "Do you remember when you measured your plants and they grew one centimeter every few days? That is steady growth." She knows students begin to understand the concept when they begin asking questions. "When you stop growing, what is the rate of change?" asks one student. Another student answers "zero." She also asks the students to remember how their families keep track of their growth, and regularly finds that most do. She measures the students' heights at different points during the year and discusses the rate of change. Only after these examples are her students able to understand how to think about the slope of a line on a graph.

Change unknown problems are also challenging for students with language problems. For example, *Ronaldo had 5 stickers. His mother gave him some more. Now he has 9 stickers. How many stickers did his mother give him?* When students are taught key word strategies, they see the word *more* with this type of problem, and tend to add the 5 to the 9. Ms. Hall finds that acting out the problems with real objects increases her students' understanding. She asks one of the students to be Ronaldo, and one to be the mother, and the class counts out loud together to figure out how many stickers the mother would have to give Ronaldo to have a total of 9.

Using Their Own Language

Learning mathematical vocabulary is not the same as learning mathematics. It is crucial that students are able to communicate their problem-solving ideas and to interpret what they are asked. However, some very good problem solvers may be slow at language learning. Instead of imposing formal vocabulary at the beginning of a mathematics unit, it is important to accept informal language from students while using the formal terminology as part of the lesson to help them develop their vocabulary, and communicate more clearly.

Teachers find that when students are involved in the process of defining and illustrating the words, they have more connection with the terms:

1. Many teachers of regular, inclusive, and special education classes work with their students to generate definitions and examples for math words such as *factor, centimeter,* and *diagonal.* Students put the information in their math journals and teachers write them on posters that remain up in the classroom.

2. Teachers also post useful computation strategies to which students can refer. Sometimes the strategies are identified with student names, and, as the unit goes on, the teachers add a mathematical name or description. For example, for subtraction, the teacher might add to Tamika's strategy: *Count or add up from one number to the other to find distance between numbers; keep one number whole and subtract the other in parts.*

3. Comparison problems are notoriously difficult for students. Teachers develop vocabulary that is meaningful to their students to help them understand comparison situations.

Ms. Williams, in sharing the following episode, focuses on how she encourages her fourth-grade students to express ideas in their own language; she then transitions to mathematical terms, uses manipulatives and representations to build mathematical language, and in particular, encourages a student with expressive language challenges to build and articulate mathematical understanding.

Voices From the Field

Expressing Ideas and Developing Shared Vocabulary

In the following episode, I hope to demonstrate the growth of one of my students in expressing his ideas and beginning to reason mathematically.

Instead of teaching math vocabulary in isolation, I need to provide a classroom environment that allows my students to use "the ways of speaking" they bring to school, from their words, to their perspectives, to their own language. And this is where the process of negotiating meaning takes place: my students and I need to work together to understand each other's words and expressions in the context we are using them. In the process, we develop common ways of speaking and become better listeners and learn mathematical terms in a way that makes sense.

My work with Carlos, one of my students whose home language is Spanish, illustrates these ideas. Carlos was a very shy, reserved child, and seemed fearful of making mistakes. Our school is a two-way bilingual school, where it is common practice to begin the teaching of reading and writing in the child's home language, so he joined my Spanish literacy class. For math, we used mostly English but Spanish was always available to students. They could express their ideas in Spanish if they needed to or I would use it to clarify ideas if I found it was necessary to facilitate comprehension. As the first months of school passed, it became apparent to me that Carlos had receptive and expressive language difficulties and that these difficulties were not related to language dominance, as will become clear in the next sections.

Early on in the year, I noticed that after giving directions, Carlos would look at me without saying anything and followed what other students did. During class meetings, he would raise his hand eagerly to answer questions, but when it was his turn, he would stay silent. Although English language learners can go through a silent period if they are not permitted to use their native language, this wasn't the case in our school. In my class, I ask questions and encourage students to talk to one another. As time went by, all of my students, including Carlos, became more familiar with my expectations around class meetings. When it was his turn to talk, he would speak. I noticed that his sentences were not complete, and often he would talk about an idea the group discussed a few minutes earlier.

After analyzing what Carlos needed, Ms. Williams decides to slow down her pace and make manipulatives and representations readily accessible for Carlos and other students.

I wondered if Carlos was able to follow what students said. I made a conscientious effort to slow down and I asked students to always show what they meant using manipulatives or representations. This move helped Carlos and many other students. For example, when solving a problem that involved all possible ways to have 12 peas and carrots (Russell, Economopoulos, & Wittenberg, et al., 2008, pp. 148–151), students always accompanied their explanations by building all of the possible combinations with snap cubes, in addition to writing the equations on a chart. Carlos gradually became more cheerful and confident, both socially and academically, and his overall participation in classroom discussions was more on target.

Second-Grade Fall: Using Representations to Build Understanding

By the beginning of second grade, Carlos was participating fully in classroom discussions. He was able to apply concepts in a particular context but was still working on generalizing the meaning to apply more flexibly in other contexts.

The following transcript shows an interaction Ms. Williams has with Carlos as she and the children play Roll a Square, a second-grade game from the Investigations in Number, Data, and Space® curriculum (Russell et al., 2008, pp. 63–65). In the game, children build 10 columns of 10 snap cubes each to make a 10×10 square.

Each participant rolls two dotted cubes (dice). They add up the total of the dot cubes to determine the number of snap cubes they will have to add to the construction. After doing so, they draw a card with a question they have to answer. In this game, children get experience counting up to 100 by 10s (and 1s), and figuring out how far they are from 100 and 50. It was Carlos's turn. There were already 72 cubes on the table, 7 sticks of 10 and 2 cubes snapped together to build the next stick of 10. Carlos rolled 2 cubes: 3 and 1. He said 4 and got 4 cubes. He snapped them on to the two loose ones.

(Continued)

(Continued)

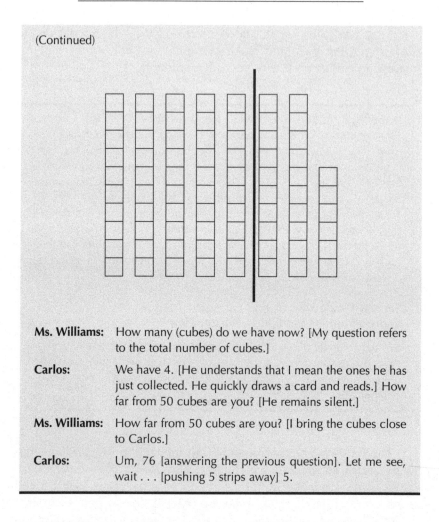

Ms. Williams:	How many (cubes) do we have now? [My question refers to the total number of cubes.]
Carlos:	We have 4. [He understands that I mean the ones he has just collected. He quickly draws a card and reads.] How far from 50 cubes are you? [He remains silent.]
Ms. Williams:	How far from 50 cubes are you? [I bring the cubes close to Carlos.]
Carlos:	Um, 76 [answering the previous question]. Let me see, wait . . . [pushing 5 strips away] 5.

Ms. Williams can immediately see that Carlos is confused. There are several numbers in play here. The amount he had rolled previously, the amount he rolled on the current turn, and the new total of cubes. Introducing *How far from 50?* is a new idea for Carlos, and perhaps an unfamiliar question. She attempts to clarify.

Ms. Williams:	Yes, this is 50 [clarifying]. So how far away did we go?
Carlos:	How far . . . 3.
Ms. Williams:	Where are the 3? Show us.

Carlos:	[He gets the 2 sticks of 10 and the one with 6 cubes.] It's 6, 10, and 10.
Ms. Williams:	But you said 6, 10, and 10. How much does this make altogether?
Carlos:	It's 76.
Ms. Williams:	No, these. How much do these make altogether? [I ask while holding in my hands the 2 strips of 10 and the 6 cubes he had separated from the 50.]
Carlos:	They make 3.

Here Carlos counts each group (10, 10, 6) as one to make a total of three groups. The other students join the conversation to try and help Carlos make sense of the problem.

Manuel:	Altogether, Carlos.
Carlos:	It's 3.
Ms. Williams:	Manuel is saying altogether.
Manuel:	And 20 and 6 is 26.
Carlos:	That's 76.
Yanira:	It's 20 [pointing at the 2 strips], 21, 22, 23, 24, 25, 26 . . . 26.
Manuel:	Carlos, just those [pointing to the 26 cubes]. Just those, just those Carlos.
Ricardo:	Not these [pointing to the 50 blocks on the table], just those [pointing to the 26 I was holding in my hand].
Carlos:	Oh! Now. He is right, because it is 20, 26.
Ms. Williams:	But why were you saying 3? Were you looking at the sticks?
Carlos:	No, but he is right because 10 plus 10 is 20 and 6 more is 26.

Ricardo's gestures help Carlos understand each total being asked for in the problem.

Ms. Williams:	So how far from 50 are we? [I place the cubes back on the table separated in 50 and 26.] Are we too far from 50 or a little far from 50?
Carlos:	A little far.
Ms. Williams:	How many cubes away from 50 are we?
Carlos:	A little far.
Ms. Williams:	How many cubes more than 50 do we have?

When Ms. Williams rewords the question, Carlos readily gives the correct answer.

Carlos: [With no hesitation] 26.

Through the conversation it became clear that Carlos was struggling. He answers 3 to the question, *How far are you from 50?* So, it seemed to me that he means three groups (10, 10, and 6). Perhaps the word *group* was not clear or the wording my question or the wording of the game card. This interaction made me think about how complex it is to understand children's interpretation of language, and how they express their thinking. These issues are even more involved with students who are functioning in two languages. In class, we had been talking about groups of 10 cubes as a unit. However, I wondered if Carlos overgeneralized the idea of a group to any configuration of cubes.

As the conversation continued, I wanted to make my question more direct, so I asked, how much does this make altogether? In spite of my best intentions, this question uncovered yet another confusion. For Carlos, the word *altogether* didn't refer to the sum of the 10, 10 and 6 cubes but to *all* the cubes that were on the table. He wasn't able to apply the word *altogether* to this specific context.

Using representations and manipulatives to "show" the concepts was also helpful to build not only Carlos's mathematical language and understanding, but also the ability to communicate mathematically for all of my students. They were invested in helping Carlos make sense of the problem. I think Carlos made progress in his understanding because the support came from the other children, not just from me.

As was often the case in Ms. Williams's classroom, the representation with snap cubes serves as the focus for the conversation. Ms. Williams continues to pay careful attention to how and when Carlos is understanding and not understanding mathematical language.

Because she encourages mathematical conversations in her classroom, she is able to ask questions that spark students to exchange ideas and help Carlos sort through his confusion.

Negotiating Meaning

By the spring of second grade, Carlos made progress in expressing mathematical ideas. This particular example, showing the progress he made in his thinking, is about story problems that involve comparison. Because story problems are often challenging, I like to introduce them by telling a story, using manipulatives for the two groups and asking students to say in their own words what they understand the story is about.

Some students initially thought that comparison problems mean how many in all, but other students explained the story context as the smaller set "catching up" to the larger set. The group came to an agreement that the second meaning is the correct one, and the expression *catching up* became part of the group's math vocabulary associated with the expression *how many more*. Although I continually use this expression, students alternate between both expressions—*how many more* and *catching up*—until they are comfortable with the mathematical language.

The approach Ms. Williams takes here is very different from the key word approach. Initially, some of the children who might have been exposed to that approach thought that how-many-more problems were the same as *all together*. However, by allowing them to express their thoughts and reason mathematically through contexts and representations, Ms. Williams's students, including Carlos, make sense of the problems for themselves.

Students solve comparison problems using addition or subtraction, so another relevant idea to discuss is the inverse relationship between addition and subtraction. The following conversation illustrates that the students are generating ideas on their own; I didn't need to ask as many questions. This is also something that happens over time.

I had given children the following problem: *I have 43 rocks and my sister has 68. How many more rocks does my sister have?* Each child worked independently and when they shared their strategies, this is what happened:

Shannon: You have 43 and your sister has 68. I took away 43, then it is 25.

Carlos: You want to catch up to get to 68.

(Continued)

(Continued)	
Shannon:	That's what I am doing, if you take away 43 from 68, then you have 25.
Carlos:	That's not catching up, that's taking away.
Ms. Williams:	That's an excellent comment. Carlos can you explain what you think using the cubes?
Shannon:	[Sounding annoyed] He is right but I am doing it a different way!
Ms. Williams:	I know; there are different ways of doing this, and what we are trying to do now is understand his way and compare it with yours.

Ms. Williams focuses the conversation on the difference between the strategies that Carlos and Shannon have used.

Carlos:	You have to catch up to 68. That means you need to put like more 10s: 53, then 63, 64, 65, 66, 67, and 68 . . . I did it like that.
Ms. Williams:	And how many did you put?
Carlos:	I put like 20 and . . . 25.
Shannon:	He did the same way as me.
Carlos:	No. But not like that.
Ms. Williams:	Carlos said if I have 43 I put 25 more, because I have 43, 53, 63, 64, 65, 66, 67, 68. That's 10 plus 10 plus 5 equals 25. He put 25 more and got to 68.
Shannon:	And I did 68 take away 43. You need 25 more rocks.
Ms. Williams:	So you started with 68 and you thought Ana doesn't have 68, she only has 43, so you took 43 away from 68, to find out how many are left?
Ricardo:	Yep, 25.
Carlos:	He is making it into opposites.
Ricardo:	That's just what I was going to say. Shannon took away and he plused [illustrates what each boy did with the cubes]. She took away 25, and she likes to do take-aways. And he likes to do pluses and he plused.

This is another example of how a student enters the conversation and takes responsibility for comparing student's ideas. In the next passage, Ms. Williams continues to monitor Carlos's understanding.

Ms. Williams: Carlos, what did you mean by opposite?

Carlos: [He goes to the board and shows how the numbers appear in the two different equations: $68 - 43 = 25$ and $43 + 25 = 68$.] I meant 68, 68, that's the first, that's the last one, 43 is the first one, 43 is the second one, 25 is the last, and 25 is the second.

Ms. Williams: So you think that all the numbers are in different order and that's why you think it is the opposite?

Carlos: They are in different order.

Ms. Williams: So Shannon starts with a big number, 68 rocks and takes away the ones that I have and then she knows how many more I need.

Shannon: Yes, exactly.

Ms. Williams: But Carlos starts with the ones I have and . . .

Carlos: Add 25. Or maybe, maybe you found 25 more rocks.

In this conversation, the children are re-negotiating the meaning of *take away* and *catch up*. Up to this point, they had not understood how they were interconnected. This conversation brings up first the confusions of connecting them, and through this confusion, they began to elaborate how they might be interrelated as Carlos said that take away and catch up are opposites. Although I knew this was the first of several conversations that would clarify these ideas further, I was pleased that the students were used to and eager to share their ideas and listen to one another. I was especially pleased that Carlos was an integral part of the conversation.

Instead of assuming that her students' language difficulties get in the way of learning mathematics, Ms. Williams allows her students to express themselves with the language that is familiar to them, often with the use of manipulatives and representations to focus their talk. The conversation and exchange of ideas she encourages helps all of her students, including Carlos, build understanding as their ideas become refined and clarified. As their understanding grows, they are then able to use mathematical vocabulary because it makes sense.

Using Representations

As in the example from Ms. Williams, teachers encourage the use of representations and manipulatives as the focus of discussion. Students can "explain" their strategies using number sentences and visual representations. Ricardo gestured to Carlos about the groups of cubes under discussion. The models and representations can also show the action of the operation, especially when mediated by language, gestures, and so on. For example, a student can illustrate multiplication by showing how to make equal jumps on a number line.

Technology can support students in expressing their mathematical ideas in a variety of ways other than the traditional approach of writing equations. Some new devices provide students with a larger range of input possibilities. The Livescribe smartpen (www.livescribe.com/en-us/), for example, captures both drawing and sound and allows students to express their understanding in a combination of written and oral modes. Tablet computers, for example, the Lenovo X201 (http://shop.lenovo.com/us/notebooks/thinkpad/x-series-tablet), provide students with a powerful drawing tool that can include a large number of colors, different width pens, and highlighters.

A representation can clarify the question being asked. Teachers of students with language difficulties often find that they have to pay careful attention to how they ask questions and what they might need in the way of representations or visual models to help students understand the question at hand. In the following interaction, Ms. Corbin is puzzled when Stefan answers her question incorrectly. Her drawings appear to help him make sense of what she is asking.

Ms. Corbin's class is acting out a "store." Stefan has $3.50 and needed $5.00 to purchase his item.

Ms. Corbin: Stefan, do you have enough?

Stefan: I need $1.50 in order to get $5.00.

Here it is clear that Stefan is able to figure out the difference between $3.50 and $5.00.

Ms. Corbin: So how many quarters do you need to get to $1.50?

When he doesn't reply, she rephrases:

Ms. Corbin: Well, $1.50 is the same as how many quarters?

Stefan: It's 2 quarters?

Ms. Corbin: Just 2?

> Ms. Corbin writes that he needs $1.50.
> She starts drawing quarters, and he tells her to stop at 6.
> She knows that Stefan can count by 25s.

Ms. Corbin: So how many more quarters do you need?

Stefan: I need 6.

So Ms. Corbin is left to ponder, was it her question that Stefan didn't understand? Or does he need pictures of the quarters to activate his knowledge of counting by 25s?

Consider the following dialogue in which a student is confused about which is bigger $\frac{1}{2}$ or $\frac{2}{3}$. The teacher has made pattern blocks readily available to her students.

Ms. Vega: So Franco, is there a little confusion for you? Is there anything in the room that might help you figure out the fractions?

Maria: He can look at the pattern block shapes and the equal pieces.

Ms. Vega: You are making a connection with pattern blocks. Can I show $\frac{1}{2}$ with pattern blocks? What shape is half?

Ms. Vega builds on a connection made by a student to highlight the particular fractional relationships.

Class: Trapezoid.

Ms. Vega: Which one is $\frac{1}{6}$ going to be?

Class: Triangle.

Ms. Vega: [To Franco] Do I need just 1?

Franco: No, 5 more.

Ms. Vega: You knew that? So when I look at this what can I say?

Franco: Well, 3 triangles is $\frac{1}{2}$ and 4 triangles is $\frac{4}{6}$.

Ms. Vega: If I put 4 triangles together, is there another way to show it? Is there a way to make this shape with something else?

Ms. Vega puts 2 rhombuses together; they match: Lots of hands are raised, and students are clapping when they match.

Ms. Vega: Those 2 rhombuses, they're each called?

Emilio: They're called $\frac{2}{3}$.

Ms. Vega: You thought 2/3 less than $\frac{1}{2}$ [holding up the trapezoid and 2 rhombuses]. Which is more?

Emilio shows that $\frac{1}{2}$ is less than $\frac{2}{3}$ by holding the trapezoid next to the 2 rhombuses.

Maria: If we add 1 triangle to the trapezoid, it is equal to the 1 rhombuses, $\frac{2}{3}$.

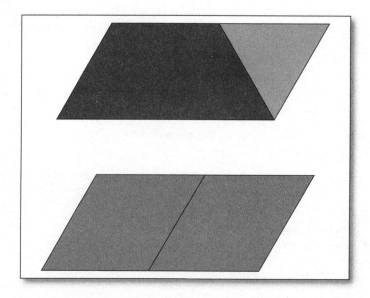

Ms. Vega: Yes, that's great. You both made a connection with something you've touched. That's why we use pattern blocks.

In Ms. Vega's class, it is acceptable to express confusion. She encourages the students to use familiar models and involves the class in figuring out the fractional relationships. However, she knows when it is helpful to provide information, for example, showing the class the rhombus to encourage them to see relationships among the shapes. The students begin to incorporate the names of the shapes as they become familiar with the relationships.

Providing Structures

Sometimes teachers find that students with expressive language problems have a difficult time entering into a conversation. After carefully considering her students' needs, Ms. Tran decided to provide her students with a structure called *sentence frames* during a unit on fractions.

Voices From the Field

Supporting Students' Language

As I was to venture into my second year of teaching fifth grade, I began to sort through my students' records to find that many of my students were English language learners and one fifth of my students had individual education plans (IEPs). I noticed that although some of my students had different learning needs, for example Nina and Kendall were on IEPs and Lourdes was learning English, what I wanted to focus on was supporting them to understand oral directions and stating their math solutions and thoughts.

I decided to start the year with putting these students' labels in the back burner of my brain and to just observe them as individual learners. My goal in the beginning of the year is to help students socialize and learn while building our classroom community of supportive, collaborative learners. In my classroom, we build math understanding together, through talk in large and small groups. Through my observations and through listening to students, I began to plan ways to build the strengths of my students, particularly Nina, Kendall, and Lourdes, who needed help in communicating mathematically.

Planning Support Systems for Language

When Nina, Kendall, and Lourdes were given time to partner talk, they would always tell the other person to go first or would just not state anything at all. Also, when I asked, "what do you know" when guiding them through math problems, in most cases, they would just stare and say nothing. As I gave the students wait time, I would count up to 30 or 40 in my head to make sure I was giving enough think time. Even when they occasionally did respond, the tone of the phrase came as a question rather than a statement of knowledge, revealing they were not confident in their response.

Although students are grouped into four and desks face toward each other to facilitate and support as much academic and casual discussion as possible, these particular students did not take the opportunity to talk as much as their peers did.

(Continued)

(Continued)

After observing Nina, Kendall, and Lourdes for some time and thinking about their learning needs, I knew that as we began our fractions unit I needed to develop particular structures to support these students so that they could access the mathematical ideas. To encourage more language from everyone—including Nina, Lourdes, and Kendall—I decided to think of talk formats such as sentence frames that would help them engage within the lesson and math games. Chapin, O'Connor, and Anderson (2009) state that talk formats support students' learning by helping students organize conversation and creating respectful equal access norms of participation within the classroom.

Supporting All Learners With Sentence Frames

As a way to support all learners within my classroom, I decided to introduce these sentence frames as whole-class norms to use during math lessons. Some sentence frames that I introduced were:

I know that _____ because _____.

I know that _____ so I did _____.

I find sentence frames useful because while they provide a structure, they are also open-ended so that the student's thinking is the major component. I introduced these through chorally stating it with the students for several days, and then using them within several examples during math and other subjects to show these sentences within context. Students were inclined to use these sentence frames and it began to become second nature as they talked about solutions and how they started solving problems during math. Lourdes, Nina, and Kendall latched on to these sentence frames and this helped them speak more during math lessons to their partners and me. However, I noticed that they needed additional support during language dependent lessons such as math games. Although students used the sentence frames during the share portion of the lesson, they needed more support through sentence frames when they were expected to interact with each other in small groups during math games.

Ms. Tran introduces a structure that is helpful for all of her students, but particularly works well for Lourdes, Nina, and Kendall. From her research, she has learned that talk formats such as sentence frames facilitate students' ability to organize and express their thoughts, and reinforce classroom norms of listening and respecting each other.

As the students began the In Between game (Russell et al., 2008, pp. 73–75) to compare and order fractions it was important for students to discuss how they knew where to place fraction cards. The goal of the game is to order fractions from least to greatest using landmark fractions of zero, half, and one whole. Students are expected to place the landmark fractions down first (zero, half, and a whole) and then draw six fraction cards each and take turns placing the fraction cards in the correct order. At the end of the fractions unit, students are expected to compare and order fractions using landmark fractions and their knowledge of fraction equivalents.

In order to support Lourdes, Nina, and Kendall's language during games such as In Between, I decided to teach them sentence frames that would help them use math vocabulary such as *greater than, less than,* and emphasizing the *th* in stating fractions. The sentence frames that I taught them were as follows:

I placed _____ between _____ and _____. I know that _____ is (greater than/less than/equal to) _____ because _____.

Again, I used the same format of chorally practicing the sentence, practicing and teaching the words *greater than, less than,* and *equal,* and practicing these words in math contexts that they knew such as whole numbers of 0 to 100 while looking at the hundreds chart. Again, the power of this structure is that students begin by specifically stating what they did, but then they need to justify their thinking.

Ms. Tran has clear mathematical goals in mind for her students, and she plans the sentence frames she gave them accordingly. She also establishes a routine for introducing the sentence frames so that the students know what to do when using them with partners.

All three students used the sentence frame with their partners as the whole class played the game. This sentence frame forced the whole class to talk about what they knew about the fractions while ordering the fractions on an invisible number line. As I circled during math workshop, I watched an observed how Nina and Kendall were interacting and talking about what they knew. During the unit, students learn different ways to compare fractions based on percent and fraction equivalents. Students are expected to know most common percent equivalents as well as fraction equivalents.

(Continued)

(Continued)

Discussions included concepts such as if the numerator is half the denominator than it is equal to half, and establishing the relationship between eighths and fourths, thirds and sixths, fifths and tenths. As students play the In Between game, they are expected to use these comparisons to rationalize their answers. These were the cards they had started the game with:

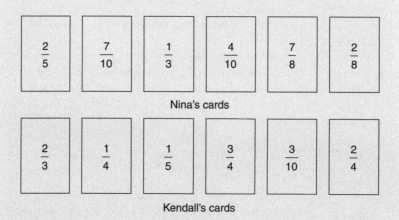

Nina's cards

Kendall's cards

During the game, this is what I observed:

Kendall: I placed $\frac{2}{3}$ between $\frac{1}{4}$ and $\frac{2}{5}$. I know that $\frac{2}{3}$ is greater than $\frac{1}{4}$ because $\frac{1}{3}$ is equal to $33\frac{1}{3}$ percent.

Nina: [Looking confused] The card isn't $\frac{1}{3}$. It's $\frac{2}{3}$ so $\frac{2}{3}$ is equal to . . . [long pause].

Kendall: [Yelling] 66!

Nina: Hmm . . . it's $33\frac{1}{3}$ percent [emphasizing the $\frac{1}{3}$ part]. So $66\frac{2}{3}$ percent!

Kendall: [Immediately grabbing the card and moves it] I placed $\frac{2}{3}$ between $\frac{1}{2}$ and $\frac{3}{4}$. I know that $\frac{2}{3}$ is greater than $\frac{1}{2}$ because it is equal to 66 and $\frac{2}{3}$.

All three students used the sentence frames during the In Between math game and this supported their partner interaction and academic language development. They used relevant math vocabulary and talked about their reasoning.

Thinking About Sentence Frames

The use of a talk format such as sentence frames helped not only Nina, Kendall, and Lourdes, but also the whole class. These sentence frames create a norm that all students are expected to have a voice during discussions. It is important for students to discuss with each other what they already knew and why they made certain moves as they played the game. These discussions enabled students to discuss their understandings and more importantly correct each other's misunderstandings. It is important to guide students in using sentence frames and to use them as a starting point to get in the habit of expressing ideas and mathematical thinking as they develop a repertoire of mathematical language.

Ms. Tran's example illustrates a structure that can be adapted according to the mathematical objectives of the lesson. The sentence frames do exactly what they say, provide a framework, but are open-ended so that they encourage students to express their thinking. They are not designed to be "fill in the blank" to teach vocabulary, as in the cloze technique in literacy (see LD Online, n.d.). Sentence frames can be used in various ways in math class. For example, in learning the characteristics of geometric shapes, a sentence frame might be,

This shape has _____, _____, and _____.
This shape has _____, _____, and _____;
therefore, it is a polygon.

At a more beginning level, the frames might be,

This is a _____. It is/has _____.
This is not a _____. It is/has _____.

In learning sequencing, a sentence frame might be,

First, I did _____. Then I did _____. (See
Bresser, Melanese, & Sphar, 2009).

As Ms. Tran states, by introducing sentence frames to the class, she makes the point that all students are expected to share ideas.

Summary

Instead of constricting students' language by "teaching" formal vocabulary at the beginning of a math unit, these teachers encourage students to express ideas in language that is familiar to them; connect new concepts to students' prior experiences; offer a variety of

manipulatives and representations; and provide open-ended structures, such as sentence frames to facilitate the expression of ideas.
Knowing that they can use familiar language increases students'
confidence and builds a foundation of understanding. Once they
have that foundation, they can integrate mathematical vocabulary in
a way that makes sense to them. Not only do these strategies help the
students understand concepts, but they also provide the teachers
with more information about how their students are learning, which
in turn, helps them refine their strategies further.

3

Helping Students Build the Capacity to Attend to and Focus on Mathematical Ideas

Working with students who are active and have difficulty focusing is especially challenging in an inclusive classroom with a wide range of needs. Furthermore, there are many complexities in considering how to plan for students who seem to have problems focusing and attending. This chapter briefly lays out some of the complexities involved in considering attentional difficulties, and then provides examples and episodes concerning students who need help in focusing on mathematics ideas, no matter how they are labeled.

Diagnosis and Labeling

Students who are diagnosed with attention deficit hyperactivity disorder (ADHD) represent one of the most prevalent categories of childhood disorders (Centers for Disease Control and Prevention [CDC], 2005). Estimates vary with some studies finding that ADHD affects 5 to 7 percent of school-age children, or about 1.6 million children

(CDC, 2005; Zentall, 2006), and others estimating that 3 to 10 percent exhibit clinically significant levels of ADHD symptoms (Barkley, 2006). ADHD consists of three subtypes: (1) combined type, (2) predominantly inattentive type, and (3) predominantly hyperactive-impulsive type. Most diagnosed cases are the combined type, both inattention and hyperactivity-impulsivity (Zentall, 2006). School outcomes for this population have not been positive. Students with ADHD are often at risk for academic failure. They are more likely to receive lower grades and lower test scores, and to drop out of school, even if medication can reduce some of the core symptoms (Barkley, 2002; Dendy, 2000; Loe & Feldman, 2007; Zentall, 2005).

There is some evidence that, in mathematics, students with ADHD have difficulty with problem solving and conceptual knowledge (Zentall, Smith, Lee, & Wieczorek, 1994). Decreased ability to focus, to organize time and space, and to master computational facts might all contribute to a generally lower performance on problem solving, and these aspects must be taken into account when planning for the mathematical learning of students with attentional problems.

To be diagnosed with ADHD, a child must have six or more of the symptoms[1], have had symptoms by age 7, and be suffering impairment in two or more settings from the symptoms (usually home and school) (Currie & Stabile, 2006). ADHD is a clinical diagnosis; there is no single test, rating scale, or so forth that provides a clear indication of the condition. Multiple sources of data and assessment are recommended, including interviews with parents, teachers, and students; behavior rating scales; school records and so on (DuPaul & White, 2004). Thorough reevaluation is also highly advisable as symptoms can change or become more mild (Rabiner et al., 2010).

Yet in practice, the ADHD labeling is particularly complex. Data has also shown that there is considerable variation about the process of identification, treatment, and educational placement (Angold, Erkanli, Egger, & Costello, 2000; Harry & Klingner, 2006; Jensen & Cooper, 2002; Jensen et al., 1999). The studies on disproportionality previously referred to show ADHD to be particularly overidentified among African American students. There is also a wide variation in how teachers use behavioral rating scales for diagnosis. For example, when African American teachers used a rating scale in one study, they rated African American students with fewer problems than did the European American teachers (Reid et al., 2001; for other examples of differential diagnosis, see also Mandell et al., 2008; Currie & Stabile, 2006). Further, there is evidence that children's patterns of engagement and participation may vary according to cultural background. Studies of U.S. Mexican heritage pueblo children have shown that they make use of learning through observation of surrounding events and overhearing cultural practices An understanding of repertoires of attentional practices might contribute to an understanding of the resources and experiences children draw on to help them learn (López, Correa-Chávez, Rogoff, & Gutiérrez, 2010; Mejía-Arauz, Rogoff, Dexter, & Najafi, 2007; Mejía-Arauz, Rogoff, & Paradise, 2005).

Classroom Strategies

The complexities in making a diagnosis of ADHD, however, do not diminish the challenges teachers face in addressing the mathematics learning of students with attentional difficulties. In general, there is limited knowledge of how to design effective interventions for students with ADHD, especially in mathematics. Many of the intervention studies have been conducted in special education classrooms, despite the fact that most children with ADHD are included in general education classes (DuPaul, 2007).

Existing research findings point to general strategies that help students with attentional problems in inclusive classrooms, for example, smaller class size, peer tutoring, offering choices, and increasing physical activity (DuPaul & Stoner, 2003; Loe & Feldman, 2007). The "School Supports Checklist" (McKinley & Stormont, 2008) includes many general strategies that are considered effective. Some apply to math—for instance, assign fewer problems; use more projects, games, and other activities that encourage active responding; and ask students to repeat directions to tasks to check their understanding. However, examples of what these strategies look like in real math classes are scarce. The purpose of the teacher contributions in the following sections is to provide specific strategies in the context of real classroom teaching. The teachers give examples of how they keep discussions active and focused, provide consistent structures and routines, use formative assessment to understand how the students learn, and take advantage of the opportunities afforded by technology.

Whole-Group Discussions

Whole-group discussions in mathematics are particularly challenging for teachers in inclusive classrooms. In addition to keeping students focused, discussion in mathematics is in general not as comfortable for teachers, many of who are accustomed to traditional mathematics teaching in which they taught students one procedure; students completed worksheets to practice the procedure; the teacher corrected the problems and there were no discussions. The emphasis on sense-making requires different practices, and discussion is a major feature. To include students with attention problems in class discussions, the discussion must be brief and targeted to particular goals.

Discussion often happens at the beginning of a math class. The goal of introducing an activity to the whole class at one time is to make sure that all students understand the task, not to have an extended discussion. To keep the discussion active and appropriate for students with diverse learning needs, teachers may present the problem in several different ways, providing materials for students to work with when useful (e.g., pattern blocks, cubes, or numeral cards). They might introduce problems (and solutions) on chart paper or the blackboard, or use an overhead projector, as well as presenting orally.

In the following example, the teacher anticipates what her second-grade students might need in order to complete the day's activity successfully.

Ms. Perez explains that students will work with partners, in predetermined pairs, and take turns making designs with 8 or fewer blocks. Then, their partners will try to reproduce it without looking. Ms. Perez makes a design in front of her on the floor and places a screen so the students cannot see it. She calls on students to ask her questions about her design that can be answered with either yes or no. After they ask a few questions, she asks the students what they did in the activity. On the board, she writes the students' suggestions as "rules" for the game.

8 or fewer pieces

Ask questions that are answered yes or no.

You can place shapes on top of each other to make a 3-D design.

Ms. Perez also writes down words that students mention during the conversation: flat, standing up, looks like, north, above, on top, almost connecting, below, left, right. Ms. Perez draws two shapes on the board, one above the other. She writes "above" next to the higher one and "below" next to the lower one.

Ms. Perez: *At the end, I will ask what words were helpful. Take turns. You are going to need a tub of blocks and a basket for a divider. Today you will work with your blue list partners.* [She has made and posted several different arrangements of partners. She tells students which list to use today.]

As children settle with their partners at tables, Ms. Perez uses one pair to demonstrate to the class how to sit across from their partner and how to place the divider. The paraprofessional sits near a pair of students whom she thinks may need extra help. Ms. Perez moves around the room observing different pairs of students.

Ms. Perez uses a variety of strategies in this example. She not only demonstrates the activity including all of the materials needed and the placement of the materials, but she provides expectations for what language the students should use. She intentionally pairs the children to provide a balance of strengths and needs in each grouping. Through her modeling, she involves the whole class. She also refers back to the rules as they call out their questions. In this case, she is able to make use of a paraprofessional to give students extra support so that her explanation to the whole class is brief. Her clear but brief explanation of the task also helps the paraprofessional be aware of how to support the students.

Keeping Discussions Active

Teachers recognize the need for students to be active during discussions in a variety of ways. Providing outlets for physical activity is especially important because opportunities to exercise at recess or during physical education can be limited. In fact, studies have shown the behavior of students with ADHD improves when students have opportunities for exercise (Loe & Feldman, 2007). Children might act out story problems, or sing or chant multiplication facts. Some teachers have children work out problems on individual whiteboards as part of their introductory discussion. Not only does this keep them engaged, but it also provides a means of assessment. One teacher gives her students permission to lie on the floor at times, but she adds, "as long as you sit next to someone who doesn't distract you." Other teachers allow students who ask to sit on activity balls to keep them focused (Freed, 2011).

Sharing Strategies

During whole-group discussions, it can be especially challenging for teachers to manage their students who have difficulty paying attention when students are asked to share strategies. Yet these conversations are vital to help all students articulate their mathematical ideas; express confusions; analyze why a solution does or does not work; and compare and make connections among various ideas, representations, and solutions.

To keep the sharing focused and active, teachers use a variety of strategies. Listing one solution one after another can be hard for students to follow and is not engaging for students. Instead, teachers select students to share strategies with a clear mathematical goal in mind. They ask questions: *Who has done the problem the same way as Maria? How is Maria's way similar to Ben's way?* They involve the class: *Thumbs up if you got the same answer as Devin?* They have clear representations available that everyone can see and use if they so choose. With her class of preK students, Ms. Walker gives students alternatives if they cannot wait their turn or listen to other students' explanations while participating in a class game. For example, when the class is playing a "feely box shape game," she might say to a student, "You can explore with shapes until it's your turn." In the following episode, Mr. Daniels describes how he finds that rehearsing with students to share their strategies during discussions increases their engagement and ability to understand the mathematics.

Voices From the Field

Rehearsing

There were two students in my fifth-grade classroom this year who exhibited difficulties paying attention or focusing for a sustained period of time. However, neither exhibited disruptive behavior in the classroom. It has been my experience that students such as these are often overlooked and ignored because they are not loud and/or distracting to others.

Serena entered my classroom fairly proficient in procedural-based work. She was also successful when working on situations where context was not required. For example, she could effectively solve an expression such as $125 \div 5$ using an algorithmic procedure. However, she would have difficulty if that same expression was put into context such as *125 students going on a field trip* and they needed to be put into groups of 5. She struggled when the mathematical content required focus and required multistep and/or an explanation of her reasoning.

My other student was Ariana. Like Serena, she had trouble focusing during mathematics class. Yet, unlike Serena, she had always performed below average in most areas of mathematics, and it was quite challenging for her to access the fifth-grade content. Ariana was open to attempting mathematical problems, yet if she didn't experience success rather quickly, she tended to disengage from the mathematics.

I had worked with both students throughout the year with more traditional approaches for engaging students who lacked focus and had trouble paying attention. I spent time with each of them individually and in small groups. I routinely asked them for input during whole-group instruction. From these tactics, I had seen little growth with each student. Typically, I observed that these two students were less involved in whole-class discussions (particularly at the beginning and the end of our mathematics time). I wanted to find a way to engage them during these times.

During workshop time, the class works in small groups or individually to solve nonroutine problems that require students to explain their thinking or solve a problem in more than one way. I made it a priority to spend time with these two students, especially because these were the type of mathematical problems that caused them difficulty. I was determined not to let either of them fall through the cracks. I discovered that I could successfully engage them in the mathematics during workshop time. When one of these students was able to solve a problem in a way that I thought was worthy of sharing, I would let them know in advance that I was planning to ask them to share their thinking during the closing mathematical discussion.

For example, Ariana was working on a division situation with leftovers: *There were 74 students going canoeing in canoes that could hold 4 students each. How many boats would they need?* She modeled the problem with color tiles and came up with a representation of 17 boats and 2 extra. She discussed with me what the remainder 2 meant in this situation (2 leftover students). Ariana was able to determine that she would have to get another canoe for these 2 students. After watching Ariana solve the problem it became clear that the manipulatives proved to be key in this situation. Because the tiles represented the students, Ariana was able to see that 2 students were left out. We reviewed how she would use the color tiles to represent the students and how they would be grouped in 4s and that the 2 that were leftover represented 2 students who would not have canoes if we did not get another canoe. I had chosen this problem with small numbers for Ariana to help her understand the concept of remainders. I realized that other students were confused about remainders and would benefit from a class discussion using the tiles to represent the leftovers. Ariana and I also talked about how to use our document camera to share her findings in our closing meeting. I made sure we had this conversation so that she would feel comfortable sharing her solution in front of the class and that she would be able to focus on the mathematics, and not get caught up in the logistics of the document camera or the manipulatives.

Mr. Daniels not only makes his expectations clear to Ariana about sharing her work, he reviews with her the strategy that she used to solidify her thinking. He also uses the document camera to record her process with the manipulatives so she will not have to re-create it in front of the class. This rehearsal process helps Ariana feel comfortable and confident.

> I found that sharing with the class not only increased both students' engagement and attention, but it also allowed them time to process what they wanted to say and to adjust their work before they were called in front of the class. This sense of success carried over to our mathematics workshop time. When they experienced success in mathematics, they were more likely to participate the next day. Although I was encouraged by the positive experience Serena and Ariana had with sharing their work, I continued to search for ways to engage them and build their mathematical knowledge.

Keeping Consistent Structures

Keeping a consistent structure that provides for a variety of learning formats (large group, small group, pairs) is helpful for students who need help with paying attention. These students respond well with predictable expectations, accompanied by a range of learning environments. One structure that Ms. Robinson uses that is particularly helpful for students with attentional difficulties is allowing all her students to move about every 20 minutes. After an activity, the class might take a quick break to stretch their legs. Sometimes they might all go to get a drink of water, run the stairs (to get the blood circulating), or go on a learning walk (to see the work of younger classes and remind themselves how much they have learned in mathematics).

Ms. Gatto incorporates consistent structures into her math class. In the following episode, she explains how she incorporates the "turn-and-talk" strategy in her lessons.

Voices From the Field

Turn and Talk

I find when the math lesson structure is predictable, the students know what is expected of them and that allows more time for learning and less time for getting off task. We begin each lesson with a mini-lesson at the carpet, then move into independent or group work, and finally a group share (either at the carpet or from their seats). I try to keep the mini-lessons brief but meaningful, short enough to hold the students' interest, but long enough to communicate the key teaching points of the lesson.

A strategy that I use frequently in my class to keep discussions active is turn-and-talk. I typically use this strategy between one and three times during any given math lesson, but it does depend on the lesson for the day. Turn-and-talks may take place during a mini-lesson, independent or group work, or during group share time. When planning my week, I try to prepare ahead of time where I might include a turn-and-talk question, but they may also be decided on at the spur of the moment. How the students are doing during a lesson can change my plan. For example, I might notice many students struggling with a particular concept or skill and ask the class to discuss this. I might see a strategy that has been quite helpful and ask students to talk about why this strategy may have been useful or efficient.

If we are at the rug, students will turn and talk before sharing their thoughts on a question or concept, and they may have the opportunity to come up to the easel to point out, draw, or explain their thinking to a particular problem. The easel is at the front of the carpet meeting area where I sit. I might record a student's strategy or have a student come up and record what he or she is thinking. I tend to write down the student's name next to the idea in order to give recognition to those who have participated. I find that if his or her idea is being referred to, a student with challenging behaviors stays more focused. I make sure to refer to students with a range of abilities (pulling out ideas from children who tend to struggle, no matter how small of an idea). The students will some-times refer to the easel during their discussion and point to something from where they are sitting.

Turn-and-talks have become part of our everyday math culture, in a way, a math routine that all students know what is expected of them. By using this routine, I feel that my students are aware that they are required to be part of a productive math conversation, whether it is in a small group or within the larger group. At times, I tend to give a verbal warning to some of my students who have a difficult time staying on task. I will let them know ahead of time that they will be talking with a partner in a few minutes about a given task. This heads-up allows them to prepare ahead of time and plan for the short conversation that will occur during the turn-and-talk.

Ms. Gatto does not wait for her students with attentional difficulties to get off task. She lets them know ahead of time what will be required of them, giving them a focused goal for their work.

Setting Goals for Turn-and-Talk

When the students turn and talk, they work with one or two students at a time and are required to talk about the given task. I try to be clear about the goals for the turn-and-talk so that the students know exactly what is expected of them. I may ask them to make a prediction about something and share it with their partner, discuss why an answer or a strategy may or may not be correct, or I may even stop them during the middle of their work time to focus on one particular question in depth. If the students are working at their desks, they work in small groups, but then I may redirect their attention to one particular question or confusion that many students are having. Students stay focused longer because they work independently, talk about something specific with a partner, and then work on their own again. This helps eliminate students from working independently for too long at a time and engaging in other conversations unrelated to math. It gives them permission to talk, but with a specific task at hand. It also allows them the opportunity to ask questions and possibly clarify some confusions, especially if I haven't made it around to conference with students who might need support.

Preparing for Turn-and-Talk

In the beginning of the year, I have a conversation with the whole class about what a turn-and-talk looks like. The students, with my guidance, come up with a set of guidelines to use with this strategy. We begin by talking about mutual respect, and that everyone has valuable ideas. If someone does make a mistake, that can be a valuable learning opportunity. This year, our guidelines were: (1) Each person in the group takes a turn to talk about the question or idea. (2) When one student in the group is talking, the other student(s) is being a good listener. (3) Ask your partner(s) questions if you are unsure of what he or she is saying. (4) Stay on topic.

We practice this strategy for several days. When we practice in the beginning of the year, I may have the students talk to each other about their summer vacation, what their favorite foods are, or any nonacademic focus. I do this to have the students focus on practicing their turn-and-talk skills, not their math skills. We then do a quick share to discuss what they learned about their partner, but also how the strategy worked: Did you let your partner talk while you listened? Did you stay on topic? Did you ask any clarifying questions?

Too often, teachers make assumptions that their students will know what to do if they ask them to talk to a partner. Ms. Gatto knows that students need to be taught how to do turn-and-talk. She wants the conversations to help students learn mathematics, so the goals and procedures must be clear to all.

Berlyn: Focusing During Turn-and-Talk

Because turn-and-talks are short conversations that have immediate feedback, many of the students who find it difficult to stay on task during independent work tend to stay focused during these turn-and-talks. Each student has the opportunity to talk with a partner for a minute or so, and then we share within the larger group. Learners who struggle can express their thoughts to a partner and ask their partner clarifying questions before the whole-group discussion. Someone who is hesitant to share in front of the larger group at least has the opportunity to express their thoughts and have a point of entry into the discussion.

There is one particular child in my room this year who rarely participates during whole-group instruction. He tends to sit at the carpet with his head down, playing with his fingers. But, when he has the opportunity to discuss a particular topic with someone near him, he becomes deeply involved in the conversation. When it is time to share, he almost always has his hand up. I sat down with Berlyn one afternoon to ask him about his participation during a turn-and-talk time.

Ms. Gatto: Berlyn, I recently noticed that you do such a great job working with a partner during our turn-and-talk time. What do you like about this strategy?

Berlyn: I like talking with other kids in the class.

Ms. Gatto: What do you talk about with your partner?

Berlyn: Whatever question you ask us. [I found this response quite interesting. It's hard for me to know if Berlyn is focused or paying attention during whole group instruction.]

Ms. Gatto: Sometimes when I am teaching a lesson to the class, I notice that you have your head down and you don't usually share too much with the whole class. But, after a turn-and-talk, you almost always have your hand up. Why do you think that is?

Berlyn: I just like talking about math with another kid. Sometimes they help me get my ideas out. It's easier to talk with one person and not share with everyone else.

Ms. Gatto: But, you do share with everyone after you talk with a partner. Your math ideas can help others learn.

Berlyn: Yeah. It's just easier to tell you in front of the class after I talk about stuff with a partner.

(Continued)

(Continued)

Ms. Gatto: What did you mean when you said your partner can help you get your ideas out?

Berlyn: It's like I can say something about what we are learning, like multiplication or adding something. Then my partner can tell me if I'm right or not. It's like we work things out together.

My conversation with Berlyn was enlightening. Berlyn is a student who I always am wondering about. Is he following along? Does he understand what I am teaching? Is he listening? Will he know what to do when he returns to his seat for independent work?

Ms. Gatto realizes Berlyn deepens his mathematical understanding by talking through his confusions in a safe way with a partner.

Turn-and-Talk: Voices From My Class

Since I found my conversation with Berlyn so helpful, I wanted to know what everyone else thought about the turn-and-talk strategy. I am convinced that turn-and-talk helps Berlyn focus. Did the other students in the class find this strategy helpful?

Ms. Gatto: When we do our math learning for the day, I often ask you to turn and talk to someone about a specific idea. Do you think this is a good strategy to help you learn?

Sarah: I think it is. You always say how we can learn from each other, not always from you.

Lanaya: It's good because I like talking with my friends.

Tyshawn: Yeah, we can say things to each other and if we don't get it, our partner can help us.

Kanila: I get nervous to talk in front of the whole class, but I don't mind talking with a partner at the rug. Sometimes my partner will share after.

Juwan: I like to turn and talk because if I don't understand what you are asking, I can ask my partner about it.

Sarah: It's like what I said before, we learn from each other. I can tell my partners what I'm thinking, and then they can tell what they are thinking. Sometimes we agree and sometimes we don't.

Ms. Gatto: What do you do if you don't agree with each other's ideas?

Sarah: That depends. If you give us more time, I try to explain why I think what I do and if we have to stop then I try to share what I'm thinking to find out if it's right or not.

Berlyn: I think it's like sharing . . . I mean it can help because we can talk about it together. I think it keeps me focused. [Interesting, I didn't mention Berlyn's focus when I asked him about this strategy. But, he pointed out that it keeps him focused. Perhaps he is more aware than I thought about his difficulty staying on task.]

Glasy: I agree with Berlyn. We are sharing our ideas. Sometimes I will share with the whole class what my group talked about. It's easier that way.

Ms. Gatto: It seems that most of you think this strategy is helpful to your learning; I'm glad to hear that because we use it a lot. Glasy just mentioned something that I am curious about. She said that it is easier to share with the whole group when you get to talk with a partner or a small group. I think a few of you also mentioned something similar to that idea as well. Is anyone brave enough to talk a little more about this?

Julian: It's like when someone doesn't want to talk to the whole class. I like math so I don't mind sharing. But, I have had some partners that don't want to share. Maybe they don't feel comfortable.

Adema: I think Kanila said she gets nervous. I do too. I like math, but I don't want to make a mistake in front of everyone else.

April: Maybe it's because we don't want anyone to laugh at us, but if we only share with a small group, we can find out if we are wrong or not.

Ms. Gatto: I think that you have all been very courageous to talk about your math feelings. I think that Julian is right; some of you feel more comfortable sharing in a whole group than others and that's okay. I'm glad that turning and talking with a partner or two is helping you learn. You can learn from each other, build ideas off of one another, and ask each other questions.

The ideas that I heard from my students made me realize how important this strategy is for the students. We have been doing it since the start of the school year, but I never thought to ask how the students felt about

(Continued)

(Continued)

this type of learning. There is quite a range of students in the class. Some students are quite capable of carrying on a mathematical conversation about what they are learning, others are a bit nervous sharing with the whole group but comfortable with a partner, and others know that it is okay to ask each other questions and even disagree with each other. It's important for me to keep in mind and be sensitive to the fact that the students do worry about getting an answer wrong in front of the whole class, as stated by Adema and April.

I was impressed with the thoughtfulness of their responses. I think that the preparation we do at the beginning of the year for turn-and-talk, and the emphasis I place on listening and respecting each other, helps the children to feel comfortable to express their thoughts. This preparation is part of the process of developing our math community over time and allowing the students to feel safe and willing to take risks when sharing their ideas. Berlyn was able to verbalize his awareness of struggling to stay on task and Kanila was courageous enough to tell everyone that she gets nervous to share in front of the whole class (although she just shared that with everyone). This conversation made me realize that not only is the turn-and-talk strategy helping all of my students learn, but it is also allowing them to become more aware of how they learn best.

Ms. Gatto establishes a community in her classroom in which students understand that they are expected to share their mathematical thinking and are comfortable expressing their thoughts and vulnerabilities. Asking her students to reflect on turn-and-talk gives them a voice in building their community. Because of all of the preparation she does with turn-and-talk, students understand the purpose of the activity, and how it helps them learn. It is interesting that Berlyn is able to contribute to this discussion and to identify his problems with focusing. It is clear that through the purposefulness in her teaching, Ms. Gatto strives to include all students in learning mathematics.

Understanding What Students Know

Students who have attentional problems can find testing situations difficult. Unfortunately, sometimes it is assumed that they do not know mathematics. When Ms. Tarlow, a resource room teacher, thinks her students who have trouble focusing know more than they showed on the test, she makes the time to do an interview. In the following episode, she describes her work with Brian.

Voices From the Field

Odds and Evens

As a math resource room teacher, I work with students on individualized educational programs (IEPs) from the first grade through the fifth grade. The following episode is about one of my second-grade students. I see three second-grade students four days a week. The students had just completed an assessment that included the following problems on "odds and evens":

1. Is 15 an even or odd number? Tell or show how you know.

2. There are 14 children who want to play soccer. Can 14 children make 2 equal teams? Show how you know with words or a drawing [2].

In the second grade, students use the context of partners (groups of two) and teams (two equal groups) to investigate what makes numbers odd and even. Throughout the unit (Russell et al., 2008) students had investigated numbers that can and cannot be made into groups of two or two equal groups. One of the goals of this particular unit is that students learn that any number that can be divided into groups of two can also be divided into two equal groups (and vice versa) and that they characterize even and odd numbers as those who do or do not make groups of two (partners) and two equal groups (teams).

On his assessment, Brian, a student who understood many of the addition and subtraction concepts, but who had been diagnosed with ADHD, showed some confusion about odds and evens.

5. Is 15 an even number or an odd number? Tell or show how you know.

O $even$

$$\begin{matrix} 6\,6 & 0\,0 \\ 6 & 0\,0\,6\,6 \\ 0 & 6\,6 \\ & 6\,6 \\ & 0\,6 \end{matrix}$$

I decided to interview Brian, about the two problems from the assessment to gain a better understanding of how he was thinking about the problems, what he understood, and what confused him.

Ms. Tarlow decides to interview Brian to learn more about his thinking. She knows that sometimes his difficulty with focusing and his frustration with getting his thoughts down in writing impairs his performance, and perhaps masks what he knows.

During the assessment, Brian was allowed to use cubes and a 100 chart to solve any of the problems. He chose to use cubes, and when I interviewed him he had cubes at his disposal. I find that using manipulatives or 100 charts improve his ability to stay with the task.

We began by working on the following question:

Is 15 an even or odd number? Tell or show how you know.

Ms. Tarlow: Can you explain to me how you determined that 15 was an even number? [showing him the work he did on the assessment]

Brian: I don't know. The test was so long ago. I don't remember.

Ms. Tarlow: How about you just try to explain to me what you did. Maybe, by talking about the problem you will be able to remember.

Brian: I feel like I have been on vacation. I don't know what even or odd is. [While he is talking, he makes a tower with 15 cubes.]

Ms. Tarlow: Okay, let's start with the second question; maybe that will help you remember what you did for the first one.

There are 14 children who want to play soccer. Can 14 children make two equal teams? Show how you know with words or a drawing.

I read Brian the question and did not show him the work he did on the assessment, but rather just let him use the cubes to figure the problem out like he was solving it for the first time. That way, he couldn't become frustrated that he didn't remember how he solved the problem and could focus on the problem itself.

Brian took 14 cubes and made them into a tower. He then broke the tower of 14 cubes into pairs. I wasn't sure if he had processed the question when I read it to him the first time because he had created partners and not equal teams, so I read him the question again.

Ms. Tarlow: There are 14 children who want to play soccer. Can 14 children make two equal teams?

Without saying anything, Brian put his pairs of cubes into 2 groups. In one group there were 4 pairs of cubes and in the other group there were 3 pairs of cubes.

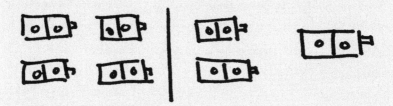

Brian: What does equal mean?

Ms. Tarlow: What do you think equal means?

I thought that Brian was asking himself this question, to give himself time to process and come up with a way to proceed.

Again, without responding, Brian created two groups, this time 3 pairs in one group and 3 pairs in another group with 1 pair leftover at the bottom.

Ms. Tarlow: How many cubes are in each group? Are your groups equal?

Brian: There are 3 cubes.

Ms. Tarlow: Okay, 3 cubes? [Was he seeing each pair of cubes as one student, I wondered?]

Brian: No, 6 cubes, and 3 partners.

Ms. Tarlow: What about these cubes down here?

Brian: If I add them to one of these groups, then one will be more.

I wasn't quite sure what to ask Brian at this point. I wondered why he had all the cubes in partners? Was he seeing each cube as 1 person or was he seeing the group of 2 cubes as 1 person? Or, did he think that his teams had to be made up of partners? Was he having trouble separating strategies, and instead was relying on what we learned about with partners and teams? What question could I ask to get at his thinking without leading him to do something with the leftover partner?

Ms. Tarlow: Why did you put all your cubes together in groups of 2?

Brian: I'm not sure.

Again I was pleased that Brian was able to verbalize his confusion, an improvement from the frustration he tended to show.

(Continued)

(Continued)

Brian then slid the leftover pair back up to join one of his groups. So now, one team had 8 students (4 pairs) and one team had 6 (3 pairs) of students. I wondered if Brian saw this? Did he think he now had equal teams?

Ms. Tarlow: How many students are on this team?

Brian: There are 6.

Ms. Tarlow: How many students are on this team?

Brian: They're not equal, 8.

Ms. Tarlow: So is there anything you can do to make them equal?

Brian: I could break these cubes apart and put 1 cube over here and 1 over here . . .

We then returned to the question about whether 15 was an even or an odd number. Brian started putting the cubes into groups of 2.

Brian: Even means they have to be together.

Ms. Tarlow: What do you mean by that?

Brian: So, 15 is odd because it's not a group of 2. Even means the cubes have to be together.

Now, I felt like we were getting somewhere. Although I was a little unsure of Brian's definition of even, I was feeling like the work we had done with even and odd numbers was coming back to him. I was especially pleased that he was able to express his thoughts.

By spending this time one on one with Brian, Ms. Tarlow not only understands better what Brian knows and what he is unsure about, but she also sees what helps him approach the task. He uses the cubes appropriately, and he is also able to verbalize, both when he is unsure and what he understands.

Brian had settled in and was feeling more comfortable with solving the problems and his own capability. I decided to ask him a new question, one that he had not done on his test to see if he could apply the work he had just done with creating teams and partners.

Ms. Tarlow: Is 23 an even or odd number? [Brian counts out 23 dinosaurs.]

Ms. Tarlow: How could you prove to me that 23 is an even or an odd number?

Brian: Put them in groups of 2 [Brian puts the dinosaurs in groups of 2.] Odd, there is a T-Rex that is leftover.

Ms. Tarlow: Is there another way that you could prove to me that 23 is odd?

Brian: I could make teams. Maybe I will have to break the head off of one dinosaur to have even teams.

I thought it was a positive sign about Brian's comfort level that he could make a joke.

Brian made one team and then made the other team. I would have expected that he would place one dinosaur on one team and then one on the other team because that is what we had done in class. When he finished making his two teams, they looked like the picture that follows. He counted 9 dinosaurs on one team and 14 dinosaurs on the other team.

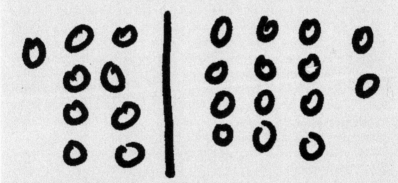

Ms. Tarlow: Are the teams equal?

Brian took one dinosaur from the team with more dinosaurs and put one dinosaur on the other team. Now one team had 10 and the other team had 13. He did this process again. So, one team had 11 dinosaurs and one team had 12 dinosaurs.

Brian: Hmmm . . . What can I do to make the teams even? If I switch one of these, then this team would be 12 and this team would be 11. I can't add another dinosaur because that's more than 23.

Again, this is another example of Brian's verbalizing out loud to help him come up with a strategy.

Ms. Tarlow: So what does that tell us?

Brian: That it's odd [referring to the number 23] because you cannot make equal teams.

Ms. Tarlow: How about if we just take this extra dinosaur away?

Brian: No, we can't do that either because then you would only have 22 dinosaurs.

My Surprises and What I Learned

Although I was pleased that Brian solved the two problems and knew that you cannot change the total number in a problem to make 2 equal groups, I realized that Brian will continue to need support to characterize even and odd numbers as those that do or do not make groups of 2 (partners) and 2 equal groups (teams).

Through interviewing him, I learned what helps him. Working with actual people or objects is a strategy that helped Brian focus and make sense. Because I had such a small group in the resource room, I could not model the "team" situations with students. I planned to talk to his teacher about doing this activity when the two children I see are in class so that they can have that experience with making teams with actual people. I try to communicate regularly with the teachers and give them each a chart of the students' strengths and weaknesses for each mathematics unit. Giving Brian opportunities to verbalize his thoughts before he attempts to write is another important strategy for Brian that I have talked with his teacher about as well.

My class has routines that give him a sense of comfort and stability. I make the materials for the day readily available and have a "do it now" problem that they do as soon as they come in. One of our structures is working with partners. I often paired him with a student who has trouble expressing her thoughts, and being able to help her gives him confidence. The anchor charts we post in the room with the various strategies helped remind him of the strategies we used to solve problems.

Brian has many strengths in math. For example, he solved number string problems using combinations of 10, can identify 10s and 1s in a two-digit number, could use a number line to add and subtract, and could visualize and make sense of a story problem about addition and subtraction.

My job with Brian was to help him harness his strengths that will help him focus on whatever activity we are doing, so that he can reason mathematically. In the case of odds and evens, through interviewing him, I was able to encourage him to use his ability to visualize to help him become more proficient with these concepts.

Ms. Tarlow makes the decision to interview Brian so she can better understand his thinking. Knowing that he has many mathematical strengths, she wants to explore what is confusing him. Through use of cubes and drawings, he is able to solve the problems and stay on task. Another strength she observes is how he verbalizes his thinking to himself as a way to help him stay focused and become aware of how he learns best. Ms. Tarlow also has routines and structures in place in her resource room to help Brian and her other students feel secure and capable of doing mathematics. She thinks carefully about pairing him so that he and the other student can work productively. In her future work with Brian, Ms. Tarlow will use this information about his strengths to plan strategies to help him in her classroom and to inform his regular classroom teacher so that he can begin to experience success in that setting as well. Often, because of the behavior problems that can result when students with attentional problems are in a large group, their strengths are overlooked.

Using Technology

Schools are increasingly using technology. It is undeniable that students growing up in a technology-rich environment find that learning through technology can be appealing. Teachers find the flexibility of many of the devices attractive. Yet like any other tool, technology can be used effectively to promote mathematical learning or it can be another means of teaching procedurally, and not focusing on understanding. To use technology well, teachers need to consider their mathematical goals and pedagogy as well as the learning needs of the students.

In the following episode, Mr. Daniels describes technology interventions he implemented with two of his students.

Voices From the Field

Engaging Students With the SMART Board

In the last few years, I have become familiar with the SMART Board and find that it is an effective tool in whole-group settings. During our whole-class discussions, I am able to capitalize on the interactive nature of the product and capture my students' attention. Many times the students will give me instructions on how to manipulate an object on the interactive

(Continued)

(Continued)

board, but most often a student or students will come to the board during these discussions. This allows for more engagement on their part and they are able to connect with the content. I also found myself calling on a larger variety of students, particularly because my students who typically struggle with focus were more apt to stay on task when they were up and moving to the SMART Board.

While using the SMART Board with the whole class, Mr. Daniels considers how to support the learning of the two students with attentional difficulties whom he discussed previously.

I was optimistic about how the SMART Board might potentially impact Ariana and Serena. I first decided to try the SMART Board with Serena during our work time with a game. Games are a regular feature of my mathematics class because I find that the student collaboration generally engages my students. The nature of these games allowed students to think strategically. The students were working on the game Roll Around the Clock[3]. Students were playing the paper version and when I saw that Serena's attention was starting to wander, I asked her to come up and play the game with me on the SMART Board. She was able to roll the fraction dice by simply touching a spot on the screen, and then was able to shade in the portion of the clock using the pen tool on the SMART Board. She was hooked. That day, Serena was not only able to become quite proficient at the game, but was able to become quite proficient with many common fraction equations such as $\frac{1}{2} + \frac{1}{4} = \frac{3}{4}$. It became clear that as she was able to manipulate the portions of the clock that were shaded in, the mathematical ideas became more clear and concise for Serena. With the technology of the board, it was much easier for Serena to manipulate the shaded-in portions of the clock and this would not have been possible on the paper version of the game. She was also more willing to shade in portions because of the flexibility and the ease of correcting mistakes.

Here Mr. Daniels describes the features of technology that help the mathematics become accessible to Serena. Not only is the novelty of the SMART Board appealing, but her ability to manipulate the clock also facilitates her understanding of the fractions.

Interacting With the SMART Board

I tried to find an entry point for Ariana, who was struggling with identifying and comparing fractions and not able to play the games that the other children were playing. Consequently, her focus was even less than normal. She was off task throughout the sessions and could not stay focused even when redirected. I was able to assess that Ariana was especially having a hard time determining the relative size of fractions, as she was stuck on thinking that if the numbers in the numerator and denominator were greater than the other fraction, then that one would be the greater fraction. Our state standards require students to add and subtract fractions and it was quite difficult for her to determine if she had a correct answer when she could not identify the size of the fraction. (In other words, she thought that $\frac{4}{7}$ was greater than $\frac{3}{4}$.)

I decided to build a number line in the SMART Notebook software that began with 0 and ended with 1. There were three marks equidistant to split the line into fourths.

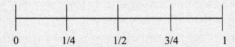

I asked Ariana to come up to the SMART Board while the remainder of the class worked during the work time. I asked Ariana to identify what fractions those three lines could represent. She was able to label them with $\frac{1}{4}$, $\frac{1}{2}$, and $\frac{3}{4}$. I decided to ask her to name some fractions that would be between 0 and $\frac{1}{4}$. She was able to tell me $\frac{1}{8}$ and $\frac{1}{10}$. I asked her if $\frac{1}{8}$ would be closer to $\frac{1}{4}$ or 0. She was quiet for quite some time. Eventually she said, "I'm not sure." So I asked her if there was anything she could do with the line to help her. She asked, "Can we make it bigger?" I hadn't considered this at all, and in fact questioned whether or not Ariana would have ever thought to enlarge a pencil-and-paper number line. However, she had seen, through another use of the board, that we had the ability to zoom in or enlarge any item on the board. With the SMART technology number line, we were able to enlarge the line so that we could only see 0 to $\frac{1}{2}$.

She thought for a while and then made the line small again, and I almost began to think she was just "playing" with the SMART Board.

(Continued)

(Continued)

She then stopped and said, "Well, if take the original line and break it into 8 equal pieces, then I think it would break each of these fourths in half." She marked approximately where each of those went. And, she enlarged the number line again so that she could just see the 0 to $\frac{1}{4}$ portion of the line. She responded with "it is exactly halfway between the zero and the one-fourth. I can see that now."

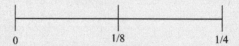

I was excited about the possibilities for her now. So, I left her with a number of other fractions and asked her to place them on the number line and be ready to explain her reasoning. Much to my delight, she got many of them correct.

Mr. Daniels observes that Ariana is able to manipulate the number line to create a new representation. The flexibility of the tool allows her to focus on a particular segment of the number line and helps her grasp the relative size of the fractions. By using the new representation to make sense of the question, she is able to reason successfully and feels prepared to articulate her ideas.

Working with students—such as Ariana and Serena, who need help focusing—is challenging, especially when they need help accessing mathematical concepts. By observing and assessing these students frequently, I was able to develop strategies that were successful in improving their mathematical knowledge. Both students showed an increase in attention and focus during the use of interactive whiteboard, gained confidence, felt more a part of the mathematical community, and were able to become more mathematically competent with fractions. At our end of the unit assessment on fractions, both Serena and Ariana scored slightly above the class average (82 percent) with an 84 percent and 83 percent, respectively.

As with any tool, the SMART Board needs to be used purposefully to be effective. Mr. Daniels has clear goals and plans for how to use this technology with Serena and Ariana. He understands their strengths and what they need to work on during the fractions unit. With this knowledge, he is able to take advantage of the software features to enhance the students' mathematical understanding. His findings corroborate other classroom examples. In a study published in *Teaching Children Mathematics,* the researchers found the virtual fractions applet they were using minimized the physical manipulation so that the students could focus more on mathematical processes and relationships among the equivalent fractions (Suh, Johnston, & Douds, 2008).

Summary

The intent of this chapter has been to contextualize the issue of attention and focusing, and to describe strategies teachers use to help their students who struggle with focus and attention, for whatever reason, learn mathematics. The label "ADHD," now among the most prevalent categories of childhood disorders, usually does not help teachers on a daily basis to teach their students. Diagnosis and use of the labels can vary by geographic region, by ethnic groups, or by gender. There are cultural variations in how students pay attention, and environmental constraints (e.g., focus on standardized testing and lack of physical education programs) that contribute to students being labeled ADHD.

The examples in the chapter show teachers who expect all of their students to participate in the mathematics classroom and seek ways to keep their students who struggle with attention part of mathematical discussions and activities. They maintain routines and structures so that expectations are clear, and discussions are active and targeted on the important mathematics. They give these students a heads-up that they will be calling on them to share, and in some cases, rehearse with them so that they know how to communicate their thinking to their classmates. All of these teaching moves require careful planning. Ms. Gatto considers what mathematical questions might be most valuable for a turn-and-talk. She then spends class time preparing them so that their roles and the goals of the activity are apparent to all. For the strategies to be effective, teachers need knowledge of each student's mathematical learning needs and strengths. For example, Ms. Tarlow describes how she makes the time to interview her students when she knows that they have more understanding than they demonstrated on a class assessment. Finally, these teachers are persistent. They take advantage of multiple tools, such as SMART Boards. As Mr. Daniels says, they are determined not to let these students "fall through the cracks."

Notes

1. Symptoms are outlined in the American Psychiatric Association *Diagnostic Statistical Manual of Mental Disorders* (*DSM*; 4th Edition, Revised 2000), and include for inattention: difficulty sustaining attention and listening, avoiding tasks, and being forgetful; and for hyperactivity: fidgeting, talking excessively, difficulty taking turns. This edition of the *DSM* includes the older term, *attention deficit disorder* (ADD), within the definition of the current term, *attention deficit hyperactivity disorder* (ADHD).

2. From Boston Public Schools End of Unit Assessment, Partners, Teams, and Paper Clips 2010–11.

3. In the Roll Around the Clock game (Russell et al., 2008, pp. 101–106), Students choose from 2 fraction cubes to roll, one with fractions $\frac{1}{2}$ and less, the other with fractions $\frac{1}{2}$ and greater. For whichever fraction is rolled, the student moves a game piece that far around the clock. The challenge is to get as close to 1 as possible and students are to record the equation to show the sum of all the fractions.

4

Developing Students' Cognitive Flexibility

It can be difficult to look beyond our preconceived notions of disabilities and how they impact student growth. Often teachers, including myself, are tempted to "give" students with special needs the answer, to try to remediate their misunderstandings by offering a sequential series of steps to follow. Michael (a student who lacked cognitive flexibility) challenged me to reexamine my ideas and he offered insights that helped me to appreciate the complexities of mathematical thinking. (Ms. Thompson)

Students who are diagnosed with conditions such as autism spectrum disorder (ASD), Asperger's syndrome, pervasive developmental disorder (PDD), or nonverbal learning disorder (NLD) often have impairments in comprehension and a tendency to adhere to rules and procedures without understanding (Burkhardt, 2007; Pennington, 2009). This chapter does not differentiate among these conditions (see the Glossary and Resource section for brief descriptions of each), but rather offers strategies and examples of teachers who are trying to help students with these diagnoses make sense of mathematics.

It is challenging for teachers to help students who need support with cognitive flexibility interpret word problems and develop

strategies that can be applied in more than one situation. Sometimes
these students can only focus on one aspect at a time and may perseverate, so generalizing does not come easily. Additionally, placing
them to work with their peers in groups also presents difficulties, as
their social skills may be limited as is their ability to communicate
what they know and how they make sense of others' strategies.

The tendency in schools has been to give these students procedures to memorize. However, if students are only taught rote procedures, they have nothing to fall back on and may experience
frustration if they make a mistake. Further, they will not be equipped
with the tools to make generalizations and solve problems.

When Michael first came to Ms. Thompson's class, he relied on
procedures and had no strategy to interpret a problem or to know if
it made sense. For instance, in the following work sample of a problem that requires students to list items from a book catalog that would
total approximately $100 (Russell, Tierney, Mokros, & Economopoulos,
2004, p. 65), Michael does not appear to work through a strategy. He
adds the dollar amounts, $52, and then gets confused in his calcula-
tions and lists the sum as $201, never realizing that this is way off
from the approximately $60 total of his list or the $100 total that was
the target number of the problem. In this problem, he is not able to
keep track of the numbers, compute correctly, retain focus on the
original goal of the problem, or make sense of the context.

Page	Book Title	Price	
7	Ice Age the movie Novel	$2.95	
6	Knights pack	$5.95	$11.
1	Guinness World records	$4.95	
8	Starry night pack	$5.95	$16
5	All-Star typing	$9.95	35
7	Spider-Man Poster Art	$3.95	
6	Pokemon Pop quiz #2	$1.95	$39
7	Scooby-Doo Campfire Mystery Pack	$4.95	
7	Rugrats Super Pack	$6.95	$18
7	S.S. life's a Beach and other SpongeBobisms	$4.95	52
	Total	201	

This chapter presents examples from teachers who do not focus on the students' diagnostic labels, but rather work hard to assess their students' strengths and learning needs, and provide a structure that helps them learn mathematics with understanding. In all cases, the teachers developed and communicated that they had high but reasonable expectations for their students to make sense of mathematics, and that students' dependence on procedures without knowing why or what they were doing was not acceptable.

One of the teachers, Ms. Neal, emphasizes that although these students have the tendency to be cognitively and socially inflexible, they also are quite different from each other, and have different strengths that can be developed. For example, one of her students can interpret a word problem, write an equation, and solve the problem, although he cannot explain how he does it. She works with him on explaining his ideas. Another student is excellent at visual spatial thinking and has done well in the geometry units. She is trying to apply this strength to numerical reasoning with representations such as arrays. Another student is musical and is able to see patterns. She is trying to build this strength to include understanding the structure of the number system.

Explicit Teaching

Teachers who help students become more flexible do not leave it to chance or student self-discovery. They name and make explicit for the student how this student best approaches problems, and develop those strengths into a strategy the student can use and generalize. These strategies are implemented from the very beginning of the year. For example, Ms. Thompson knows that Michael will not be able to follow directions in the whole group. She approaches him as soon as she can after the group introduction and inquires, "What is the first thing you are going to do to solve the problem?" and together they review strategies that get him started. At the same time, in continuing to assess his strengths, she notices that Michael has made gains in reading, has an uncanny sense of what is going on and is less distracted by details. She thinks about how Michael can follow the plot in a story, and decides that he will benefit from seeing the "plot" of a mathematics problem as well. She asks him, "What is happening in the problem?" or "What is it asking for?" To further help him focus on the meaning of the questions and eliminate some of the distraction, she at first removes the numbers from a problem and only puts them back after Michael has explained the problem. This process has to be repeated over time before Michael can begin to work independently.

Technology can also offer strategies to help these students focus instead of being distracted. For example, Michael and other students tend to perseverate and often repeatedly erase answers. Using a tablet computer (e.g., the Lenovo X201) can help these students as well as those with fine motor control problems. Students particularly appreciate the ability to delete their work using an eraser that leaves no marks behind, but students of all abilities have explicitly commented on how much they like being able to erase without making a mess.

With younger children who struggle with cognitive flexibility, a goal is to help them pay attention to what is being taught. When Ms. Hall asks students to compare two towers of cubes (students present and absent), Shauna perseverates on the "less" pile because she knows her friend is absent. Ms. Hall looks directly at her, holds the towers up right in front of her, and moves her hands along each tower to indicate the quantities, "Shauna, what am I saying?" She asks Shauna to touch each cube to bring her back to the problem at hand. Ms. Hall finds she is most effective when she uses gestures and exaggerated body language and keeps the pace rapid.

Even with her young students, Ms. Hall makes explicit her intention and her expectations. When one of her students, Greg, moves on to kindergarten, the teacher seeks Ms. Hall's advice because Greg is not able to follow the stories she reads. Ms. Hall suggests that she tape the stories and allow Greg to listen to them later, giving him instructions such as,

> I am asking you to listen to the stories because I want to make sure that you understand the story. The questions I ask you will help me know if you understand what is happening in the story and what the characters are doing. If you don't answer me, I don't know if you understand the story. If you get to the point that you can listen to the stories with us in the group, then you won't need the tapes any more, and you'll have more time to do other things.

Ms. Hall finds that with students like Greg, it is important to be up front with them about what you are doing and why, in clear and concise language.

Developing a Set of Limited Choices

Ms. Neal illustrates limited choices in the following scenario. She gives a word problem that asks, *I have 7 apples and I need 11; how many more do I need?* She starts by giving one of her students, Sam, three choices of equations and asking him which choice matches the problem. She chooses this strategy because she is confident that Sam already knows the "facts," that 7 + 4 is 11, 4 + 7 is 11. Without Ms. Neal's guidance, Sam solves all of the equations instead of picking one, so Ms. Neal has to review with him what *choose one* means. In doing this same structure over time, Ms. Neal finds that Sam can learn to make sense of the problem for himself. With other students, she sometimes has to rely on representations, asking the student to show the situation (using dots for 1s and sticks for 10s to avoid the student taking too much time to draw).

Language

The particular language that teachers use is very important, and has to be consistent. When students are not successful, teachers often evaluate how they have been explaining terms and concepts, and where the student's thinking might have gone awry. When one of Ms. Santiago's students can't answer the question *How many numbers are between 20 and 30 on the number line?*, she changes the question to *How many black marks are there between 20 and 30?* The student is now able to respond correctly, and can then make the connection between the "black lines" and numbers.

Using Representations and Contexts

The teachers working with students who need help with conceptual understanding find that it helps to offer a variety of ways to expose the student to a concept or strategy. Ms. Hall finds that Greg can only focus on one concept at a time, for example, either addition or subtraction. If they are working on subtraction she reminds him, "Is the answer going to be bigger or smaller?" If she asks him to give out supplies, after he has handed out some pencils, she asks, "How many pencils do you have now?"

When she teaches shapes, she takes her students on "shape walks" so they see the shape in different contexts. This strategy helps all of her students, but is particularly helpful for Greg. When he looks at pictures of shapes, he intends to focus on one aspect, for example, the angle or the side. When they go on the shape walk, she finds that he is able to take in the whole figure and remember images of the various properties of the shape. Another effective way to draw Greg's attention to the properties of shapes is to use a software program that asks students to complete shape puzzles (Clements & Sarama, 2010).

In helping Michael understand operations, Ms. Thompson focuses on visual representation and acting out the problem. The following example illustrates her work with Michael to help him understand division and develop strategies to interpret and solve problems.

Voice From the Field

Making Sense of Division in Fourth Grade

A story problem from the fourth-grade midyear test illustrates Michael's difficulty interpreting the following question:

> Forty-two fourth graders are taking a trip to the science museum. They are going to travel in vans. Each van holds 8 passengers. How many vans will they need?

Michael's answer was "5R2," which showed the correct computation answer, but he did not make sense of the problem. Not only didn't he refer to vans in his answer, the R2 is not reasonable given the context of the vans. We had been doing many problems in class using division in context, including many that had remainders, but Michael was not able to independently apply this experience to the question at hand.

To put the focus on the meaning of the problems, I asked Michael to use words, pictures, or numbers to explain how he solved a problem, and to provide his own story to fit a computation. At first, he interpreted the direction of "using words" literally, substituting words for numbers in his incorrect solution to the problem: *Thirty-five ÷ by seven = eight*. Asked to write a story to fit 35 ÷ 7, he showed his confusion: "If 7 kids in a class, 35 roses [rows] of tables, how can the teacher fit 7 kids?"

To help Michael interpret problems, I did a lot of visualizing, drawing, and acting out with him. Although, at the same time I was trying to get other students to use equations to express number relationships, moving them away from using tallies. I felt it was a critical support for Michael at this point to re-create the action of the problem through his drawings, such as tallies. Michael used drawings to differentiate between multiplication and division and match the action of the operation to the problem. Were we separating or combining? Giving out equal portions or putting them together?

Ms. Thompson here reveals her deep understanding of Michael's learning needs. Her goal is to help Michael access the mathematics, even if his strategy is not as efficient as the rest of her students at that moment.

As we neared the end of the year, I began to see progress in Michael's motivation to make sense of the problems. For example, I asked Michael to solve the following problem independently:

Some friends shared a crate of oranges equally. There were 40 oranges in the crate. There were 7 friends. How many oranges should each friend have received and how many oranges were left over?

As the work began, Michael excitedly shouted: "division" with none of the anxiety that once came over him as he tried to figure out the demands of a task.

The dialogue that ensued shows some marked changes in his thinking:

Michael: It's division!

Ms. Thompson: Great, now what do you do?

Michael: How should I set it up? If 7 goes into 4, can't do it . . . 7 goes into 40, ugh! I have no idea. I forgot how to divide. I know only the easy way.

Ms. Thompson: Then do the easy way.

Michael: Okay.

Michael drew out 40 tally marks and drew circles around 5 groups of 7. He successfully answered the question: 5 oranges each, 5 left over.

By turning to this familiar strategy, Michael was able to confidently interpret the problem, visualize the action, identify the correct operation, and (with small numbers) get the correct answer. Such an "easy way" was not efficient, but it made sense to him to illustrate the problems for himself to find a sensible solution strategy. In the fall, Michael would probably have tried the "how many goes into" way, and then given up and just added the numbers.

In this example, Ms. Thompson works on sense-making with Michael. For him, mathematics has been a series of procedures that frustrate him because he cannot remember the steps. She uses visual representations both to help him see the problem and talk through the action as he is drawing. The drawing and the talking facilitate his understanding of the mathematics.

Ms. Owens's student, Manuel, is not progressing in understanding the operation of multiplication. When she asks him to represent 7×2 with tiles, he takes out 2 groups of 7 and 2 groups of 2.

Then Manuel removes the groups of 2 and Ms. Owens asks, "How much is 7 times 1 plus 7 times 1?" Manuel replies, "7 times 7." She then inquires, "How many sevens are there?" Manuel answers, "7 plus 7." Ms. Owens follows up: "So, how many times do you have the 7?" Manuel is then able to say, "7 times 2." She then takes out another group of 7 and then another, and Manuel is able to say, "7 times 3, and 7 times 4." Ms. Owens moves on to groups of 8, and Manuel successfully represents them. However, Ms. Owens realizes he is counting by 1s, and is not yet seeing patterns. For example, he does not see that 12×8 is the double of 6×8. This gives her a place to start for the next day. She realizes that Manuel is making progress. Her use of concrete objects and the careful questions she poses are helping Manuel move ahead in his understanding of multiplication.

Selecting Strategies

Ms. Thompson works with Michael for two years, a decided advantage in that he becomes familiar and comfortable with her, the classmates, the routines, and the expectations for learning. During the second year, she is able to use her knowledge of Michael to help him develop a more reliable strategy. She knows that when he uses rote procedures, he does not apply his knowledge of place value, and usually arrives at an incorrect answer. She also realizes that he becomes confused when trying to solve problems in multiple ways. As she thinks about the times that Michael has been successful, she decides to spend a lot of time with Michael practicing how to break numbers apart and multiply by parts.

Voices From the Field

Breaking Apart Numbers

As a teacher, I felt good about the approach of taking apart the numbers since it required some understanding of the value of the digits and not just rote procedure. When Michael could uniformly break apart the numbers by place value in a structured way, he was successful, even when the numbers increased. He could explain what he was doing in a way that made sense.

Michael came to enjoy showing and explaining his problem-solving methods. To give him a little time to prepare, I would tell him that I was going to call on him next. He was able to explain how he multiplied 233×5:

Michael:	First I broke down 233 into 200, 30 and 3. Then I multiplied all of those three numbers by 5.
Ms. Thompson:	Like this, am I setting it up the way you want me to? [referring to my writing on the board]

200 × 5
30 × 5
3 × 5

Michael:	Yes. First I multiplied 200 times 5 and it equaled 1,000.
Ms. Thompson:	How did you know that? You knew that pretty quickly.
Michael:	I did this. [He shows me his paper on which he has added 200 5 times.]
Ms. Thompson:	Oh, you did repeated addition.
Michael:	Yes, I always forget to say that. Then I multiplied 30 times 5, which is obvious like 3 times 5 equals 15, which equals 150. Then 3 times 5 equals 15. Then I added it up.
Ms. Thompson:	What did you get?
Michael:	I got 1,165.

$$200 \times 5 = 1000$$
$$30 \times 5 = 150$$
$$3 \times 5 = 15$$
$$1,165$$

Ms. Thompson:	Did you double-check your work?
Michael:	In order to know that I added it up.
Ms. Thompson:	How did you do that?
Michael:	I did it the same way as before.
Ms. Thompson:	What's that called?
Michael:	Repeated addition.
Ms. Thompson:	So you double-checked it by adding 233 up 5 times, and you got the same answer?
Michael:	Yes.
Ms. Thompson:	I like your strategy of breaking up the numbers into parts you could work with. You actually used 30 times 5 and 200 times 5 to help you.

Ms. Thompson reinforces Michael's thinking by explicitly naming his steps and strategy. As fifth grade progresses, and the work becomes harder, Michael is able to apply his strategy to larger numbers. More important, this strategy provides Michael with a way to go back and check his work, step by step. If he makes an error, he is able to fix it without having to start all over again as he did when he used the standard algorithm. Doing this gives Michael a starting place to understand and organize his ideas and thoughts. He feels successful because he can apply this strategy to a variety of problems. He is developing a solid understanding of place value.

Working in Groups

Some of the strategies the teachers use to help their students who tend to be inflexible in groups were described in the first chapter. Teachers are honest with the students in the class about the challenging behaviors that the students who need support in being flexible display, but also carefully bring forth these students' strengths, as well as model their attitude of acceptance. Ms. Thompson has discussions with her class about how to help friends if they get upset. She also works with some of Michael's potential partners on specific ways to help him. Because Ms. Thompson's class has been together for two years, students themselves have learned how to redirect Michael before he becomes frustrated. Another teacher, Ms. Neal, discovers that one of her students who needs support with abstract thinking is at the same time very effective at teaching the class a strategy that he has mastered. When she knows that he understands a strategy, she asks him to explain it to the class. Although he doesn't think to ask students if they have questions or to expand on their comments, his explanations make sense, and she intervenes to comment on students' remarks or to suggest that he take questions. Seeing his strengths helps his classmates see him in a positive light, and they are therefore more likely to work with him in small groups.

Ms. Neal also chooses partners for her students and keeps them in these dyads for a month. Students can then get used to each other and find comfortable ways of working together. She finds that one of her students performs better in a group of three because the other students are not dependent on him for enriching their knowledge, and he is able to listen to them. With a group of two, this student has difficulty trying his partner's way and has a meltdown. Ms. Neal encourages him to take responsibility by giving him words to say: *I'm sorry I couldn't work with you. I just got stuck.* She also gives partners language to help them think about how to deal with the student. She suggests, for example, that they consider, *Sam is doing an unexpected behavior. My strategy is to ignore it and yours should be the same.* Over the years, she has had students who could only work with a one-on-one aide as a partner. In those cases, she works with the aide to bring out the student's thinking instead of telling him or her what to think or write.

With younger children, Ms. Hall finds that students who are assertive but patient can be effective partners for the students who need support with cognitive flexibility because they can keep the student on task. Sometimes a student who perseverates can actually be a good partner for a child who struggles with expressive language because the repetition can be helpful. Pairing or grouping these students who struggle with flexibility is easier when they are playing a game that has a "specific dialogue," for example, with a card game like War or Compare, the directions and wording are clear: *Turn over your card. I have more,* and so forth. After demonstrating the game with the whole group, using the overhead projector when relevant, she watches each pair through a few turns so she can catch any problems right away. Knowing that most pairs are on the right track also frees her to spend more time with pairs that might need more support, such as the students who can perseverate or be rigid.

Summary

The purposefulness of the teachers is evident in all of the examples in this chapter. Teachers expect all of their students to learn mathematical concepts, including their students who need support with cognitive flexibility. They think carefully about the mathematics they are teaching and how to make the mathematics accessible to these students with particular learning challenges. Whether they are considering what language to use, what representations to try, what strategies to highlight, or what student groupings to plan, they orchestrate the learning sequence for these students. Because it is essential to start with what the student already knows, and the positive work the student is already doing, they make the time to observe these students. They implement the approaches that they find successful for more than one lesson to help the students generalize. They also work to make the students part of the mathematical community. They use the thinking students engage in to assess their strengths and understanding throughout the year, as they continue to consider the students' progress and make adjustments.

5

Developing Strategies
for Students With
Memory Difficulties

She doesn't know any of her facts no matter how many times she writes them down.

Some days he seems to have learned his tables, but the next day the knowledge is gone.

How many times do teachers either make or hear statements like these? Research has shown that memory problems are common for students with learning disabilities (Reid & Lienemann, 2006). Typically, memory functioning is defined in terms of the length of time between the exposure and the recall, that is, long-term, short-term, and immediate memory. A more formal way of saying the same thing would be to refer to the three categories as remote, recent, and working memory (Neuropsychonline, n.d.). However, it is clear that memory functioning is complex, involving multiple dimensions (sensory, short-term, working, and long-term) as well as factors such as prior knowledge, motivation, and capacity.

It is also important to understand when and why students have memory difficulties. For example, is the student having difficulty remembering the steps to a procedure because the procedure was not well understood? Does the student understand a strategy, but cannot access it when attempting to solve a problem? Does the student have difficulty with visual memory because the math program has not offered enough opportunities for practice in this area?

There is no single strategy that will work for all students with memory issues (Swanson, Cooney, & McNamara, 2004). Mnemonic instruction is often featured as an aspect of teaching students with memory issues, and it has been shown to have some success (Scruggs & Mastropieri, 2000). However, this technique also has limitations. Becoming fluent with facts is a mathematical goal for students, but learning facts alone will not help them develop strong math skills and understanding. Students need to understand what the facts mean (e.g., that 7×5 is 7 groups of 5 or 5 groups of 7), what types of problems require using them, how they relate to each other (e.g., if they know 5×7, they can solve 6×7), and how they can use a fact to solve a larger or more complex problem. It can actually be easier to learn facts when tied to understanding them. Some of the problem-solving mnemonics focus on key words or signs. However, if students only think about key words they might miss what is actually happening in the problem. If students only see the word *more* as addition, they may not understand how to solve *how many more* problems. If they focus on the subtraction sign, they may not consider *counting up,* a strategy that often makes sense to students to solve subtraction problems.

Mnemonics that help students keep track of their problem-solving process can thus be useful. In Chapter 6, one of the teachers uses ESAL—equation, strategy, answer, label—but in a larger context of working with the student on making sense. Another general problem solving mnemonic is STAR:

- Search the word problem.
- Translate the words into an equation in picture form.
- Answer the problem.
- Review the solution (Access Center, 2006).

Opportunities for practice that capitalize on the student's particular strengths must be combined with students' making sense of mathematics.

Most important, exclusive reliance on prescriptive techniques can exclude students from access to meaningful mathematics. Often students who cannot master basic facts are discouraged from access to higher mathematics (Garnett, 1998; Hankes, 1996). When students' deficits remain the centerpiece of instruction, their strengths can be missed as instruction focuses on low-level isolated skills. The examples in this chapter focus on building students' memory in the context of making sense of mathematics. The teachers offer multiple opportunities for the students to practice their facts, at the same time including them in mathematics activities designed to help them understand mathematical concepts and ideas.

Building on Student Strengths

The following episode is an example of a student with memory difficulties who initially was not taught mathematics in a way he could learn. With family and teacher support that centered on building on his strengths, he eventually flourishes.

Darrell's difficulty with mathematics first becomes clear in the third grade. In school that year, he is not successful in learning the math work in class that consists of carrying, borrowing, measuring, and memorizing times tables. The class is large, and the teacher usually presents information orally. Because Darrell's parents are educators, and they observe him engaging in a variety of projects at home, they recognize that he is able to learn using models and visual representations. When they do not receive any support from the school system, they find a tutor who emphasizes kinesthetic and visual strategies, such as walking along various segments of a floor-length number line, using arrays[1], and 100 charts.

With his tutor, Darrell begins to understand the sequence and patterns of the 10s and 1s in the number system. The family can't afford unlimited tutoring so his mother learns the activities and works with him at home. Darrell's mother continues working with him at home, using some of the same models that were part of the private tutoring sessions the previous year. She works with him on putting 10s together to build a number line with interlocking cubes, telling time, and learning the value of coins. She plays games with him, such as "I'm thinking of 16 cents. What combinations of coins might there be?" She reports that by working with 100 charts and number lines, he is solidifying his understanding of number relationships and repeating patterns in the number system. However, when she asks the aide if she would use these materials in school, she is reprimanded for not trusting the school's judgment.

The following year, during fourth grade, his progress continues with a classroom teacher and special education teacher who work together as a team and who are knowledgeable about both mathematics and teaching children with a variety of needs. In the beginning of the year, most of his mathematics instruction takes place in the resource room. Because the special education group is small and the teacher is experienced and confident in her own mathematical understanding, Darrell is comfortable telling the teacher what he doesn't know and trusting her response. He has noticed that she encourages the students to be honest about what help they need and teaches them how to ask for help. Every student contributes to discussions during math class and gets regular feedback. Darrell's mother is pleased that the special education teacher uses some of the models he has learned before, such as the number line, and organizes the work so he can build on what he knows, such as learning facts in groups, for example, learning the doubles and then doubles plus one[2].

Darrell soon gains confidence and becomes a star in the group. Both the special education and classroom teacher report that he is becoming more familiar with addition facts and that he is developing and using strategies for solving computation problems. By the middle of the year, Darrell sometimes asks the special education teacher if he can participate in math in the general education classroom more often, adding, "If I need you, I'll come ask you." Both teachers are pleased to honor his request. The classroom teacher develops projects with multiple entry points and uses strategies that complement Darrell's strengths. For example, she introduces an array model for multiplication. Students can build arrays to illustrate the factors for smaller or larger numbers, according to what they need. The array model is particularly suited for Darrell's strength as a visual learner.

Unfortunately, at the beginning of his fifth-grade year, when the school implements an NCTM standards-based curriculum (*Investigations in Data, Numbers and Space,* Russell et al., 2008), Darrell's family is told, "He's not ready for this curriculum because he needs to be more automatic with the multiplication facts first." The school places him with 22 other students who were on IEPs under the supervision of teacher aides who have not been offered any professional development in math. Darrell brings home worksheet after worksheet and sadly comments, "I hate math."

During this fifth-grade year, Darrell's special education teachers emphasize automatic recall of facts to the exclusion of mathematical reasoning and understanding. They expect students to learn facts and procedures before allowing them to attempt problems in context. Because of this procedural teaching, they are not aware of Darrell's strengths in mathematics. They do not use his visual spatial sense to build his understanding, nor do they provide contexts that make sense to him.

His parents then insist that he be included in the general education math class, and Darrell's confidence increases. The curriculum actually fits with his strengths. For example, the fractions unit presents relationships among fractions, decimals, and percents, which draw on his accurate sense of proportion and ability to reason. Darrell's mother sees how the focus of the *Investigations* fractions unit on multiple representations of fractions and relationships among fractions, decimals, and percents fits with his ability to build on what he knows. For example, since $\frac{1}{5}$ of 100 is 20, $\frac{1}{5} = 20$ percent; $\frac{2}{5}$ is twice as much so $\frac{2}{5} = 40$ percent (Russell et al., 2008, p. 166). His accurate sense of proportion and his ability to reason help him mentally "see" the comparative sizes of fractions.

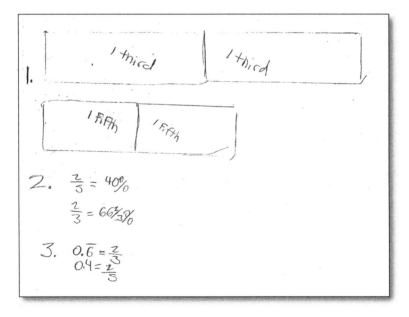

At this stage, since the school only offers Darrell limited support once he is included in the general education classroom, his mother continues to work with him at home. When she introduces a game to him, she introduces the concepts gradually, so that he can build a solid core of knowledge and incorporate new knowledge in relationship to what he already knows. For example, when they play an ordering fractions game, she limits the fractions to halves, thirds, and fourths at first, adding sixths, eighths, twelfths, and then fifths and tenths in subsequent rounds. She also continues to reinforce his understanding of the number system by playing games from earlier grade levels of the *Investigations* curriculum (Russell et al., 2008) such as Tens Go Fish, asking for the missing number to make a total of 10 (pp. 116–117); Double Compare, comparing pairs of number cards to decide which pair has the larger sum (pp. 101–102); and Capture 5, moving from one number to another on a 100 chart by adding and subtracting 10s and 1s (pp. 61–62). The extra time and practice she provides using a variety of visual models and representations solidify Darrell's knowledge of fractions of 10 and 100, providing a foundation for percents and decimals.

Sixth and Seventh Grades

In sixth and seventh grades, Darrell participates from the beginning of the year in the *Connected Mathematics Program* (Lappan et al., 2004, p. 5) as part of the general education class. He has math practice in a small group twice a week with a special education teacher, and his mother continues to support him at home. His teachers are positive about him, and in turn, Darrell likes them. He is able to make sense out of problems when given time and encouraged to use visual representations. To help with finding gain and loss in stock prices over a week's time ($1\frac{15}{16}$ on Monday, $2\frac{1}{8}$ on Tuesday, $2\frac{3}{8}$ on Wednesday, $2\frac{3}{16}$ on Thursday, and $2\frac{1}{4}$ on Friday), Darrell uses a tape measure that includes eighths, sixteenths, and thirty-seconds.

To compare fractions and solve problems in context, he often draws diagrams:

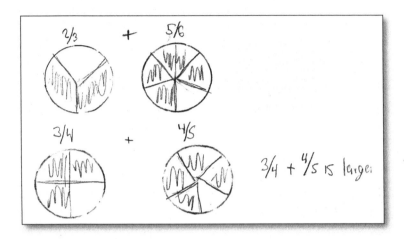

Although Darrell has gaps in his mathematical understanding, such as remembering times tables and sequencing steps for procedures for adding and subtracting fractions, he also has areas of strengths in geometry, measurement, and comparing fractions, and he has interests outside of school that motivate him to learn mathematics. After varied experiences with mathematics instruction, Darrell develops insight into how he best learns mathematics, and he becomes aware of the difference between teachers who are prepared to work with him and those who are not.

After fifth grade, a conversation between Darrell and his mother reveals his ability to reflect on his own mathematical abilities. The following excerpt from the conversation illustrates Darrell's growing ability to reflect on his school experience and to understand how he learns mathematics:

Mom: How did math class go this year?

Darrell: Okay, a little bit better than being in the separate math class.

Mom: What went well?

Darrell: Doing group work and playing games.

Mom: What could have gone better?

Darrell: It would have been better if the teacher were familiar with how the math curriculum worked. The teacher was good at doing math but not teaching math.

Mom: Describe a good math class.

Darrell: The class wouldn't be too big or too small. The kids would be at my level, or a little better than me in math so I could learn from them. Tools available when I need them, for example, fraction strips, equivalency charts, calculators; no time pressure or math drills; a good teacher.

Mom: What things in math are you good at?

Darrell: Estimating length. Reading thermometers when the lines aren't too close. Drawing humans and animals in proportion.

Mom: What things in math are hard for you?

Darrell: Some times tables. Estimating things like 424,389 divided by 28. Division.

Mom: What are ways that a good teacher could help you learn in math?

Darrell: Don't rush me. Teach me slowly step by step. Don't panic if I don't know the times tables. Don't get angry. Don't say "You should know this by now. We shouldn't spend time doing this."

The teacher knows how to do math well and teach math really well. The teacher would be able to explain things in different ways and show me strategies I can use.

Mom: What are ways you have learned to help yourself?

Darrell: On tests, I need to be calm. Like on my diving test, the final exam was a 50-question test, I got 4 wrong. I took 45 minutes longer than anyone else: the teacher knew it was okay. I thought: "Darrell, just be calm, do one at a time." Don't compare my work to others. Go really slowly and don't try to keep up with other kids.

Darrell is fortunate to have several advantages. His parents are educators who are able to support him through their knowledge of mathematics education and through resources they obtained from colleagues, and who are able to advocate for him. They also encourage him to express what works for him, what is difficult for him. His awareness of his own learning becomes a real strength. Because of his many interests outside of school, his classmates see him as capable and skilled in many areas, so his self-esteem is intact.

His positive years in school featured professionals who were knowledgeable about mathematics and teaching mathematics, who collaborated well together, and had high expectations for their students. These teachers also built on Darrell's visual and spatial strengths, instead of just focusing on his memory difficulties. When it came time to teach him facts, in addition to using a variety of representations and models, they taught him the facts in a carefully sequenced way, for example, having him work on doubles, followed by doubles plus one.

Developing a Conceptual Foundation

Starting with the early childhood years, it is essential to build a conceptual foundation before expecting young children to have automatic recall of number and shape concepts. Ms. Walker uses multiple strategies, based on the math curriculum she teaches (*Building Blocks*, Clements & Sarama, 2010). She does not expect her young preschool students to understand shapes by teaching them the definition of each shape. Instead, she provides her students multiple exposures to spatial concepts.

Ms. Walker has two students in her class who have difficulty with memory. Nina's spatial sense, regarding directions and geometric shapes, is not yet developed so she has not been able to name common shapes. When Ms. Walker begins teaching a concept, she asks herself, *What do I need to know about my students' prior knowledge to make the concept tangible?* She gives students coffee stirrers and straws to create shapes; she puts tape on the floor and has the students walk around the sides of the shapes. As they walk, she counts with them: "From here to there is one; from here to there is two; from here to there is three. How many sides is that?"

Ms. Walker makes sure that Nina is engaged in the activity and is responding to the question. If she isn't, Ms. Walker does it again, sometimes holding her hand while they walk around the shape together.

She asks her student to draw shapes in the air, giving specific instructions. For a triangle, she might say, "Go left, right, and then turn your finger. Slant up, flip your finger, and slant down." Then they might sing the shape song, emphasizing the words *one, two, three* for the sides of the triangle. In addition, Ms. Walker encourages expressive language. When a child puts his hand in the "shape feely box" and announces, "It's a square," Ms. Walker follows up with, "How do you know?" She might add, "Describe it; are the sides long or short?" If it's Nina's turn, she asks her questions that Nina can answer. "How many sides does it have? Are the lines bent or does it have straight sides?"

By exposing Nina to the shapes through a variety of approaches, she is offering Nina a solid foundation. Ms. Walker is also aware of what Nina knows, and what she wants to help her grasp. She sees the results of her approach when her students go outside and name the shapes they see in the neighborhood. Even Nina has begun to recognize squares and circles.

Another student, Andre, has not mastered many of the skills involved with counting. He is not solid with his number sequence, and does not connect number to the quantity. Again, Ms. Walker uses a variety of physical representations, such as cylinders, cubes, and counting dinosaurs. She also engages him in counting games, such as comparing dots with very small quantities, and repeating these games often. When they sing songs such as "Bingo," (in this song, students say one less letter for each round, clapping in place of saying the letter, until they are clapping five times in the last round), she holds up cards so students can visualize the number of claps for each round. While Andre cannot yet contribute to this game, he is being exposed to the zero to five sequence backward and forward. At the same time, other children in the class use what they learn from Bingo to become fluent in the combinations to five.

Ms. Walker observes Andre closely and is aware of what works for him and what causes him confusion. For example, she finds that building a sequence of numbers with a vertical staircase does not help Andre with the number sequence. Instead, she lays down one object at a time on the desk (e.g., a cube), and then he can get two, three, and four, and match them to dots. If he hesitates matching the number of cubes to the dot cards, Ms. Walker might ask, "Do you have enough?" This question seems to prompt him to recount and figure out the correct amount. Sometimes she gives an incorrect number to prompt his thinking. If he's thinking about what comes after three, she might say, "one, two, three, six."

The multiple strategies Ms. Walker uses are beneficial for all of her students, but by understanding the skills and concepts needed to develop number sense and spatial concepts, she can differentiate instruction for students like Nina and Andre, who need repetition and support to integrate these concepts, and for other students in her class who are ready for a challenge.

Grounding Skills and Concepts in Understanding

When Darrell described what he would like to see in a mathematics teacher, he was in effect describing Ms. Walker and Ms. Gatto's practice. Ms. Walker provides her young children with a strong mathematical foundation. Ms. Gatto knows that it is important for her fourth-grade students to learn their multiplication facts, but she does this in conjunction with building an understanding of the operation of multiplication. She provides many opportunities for students to work in small groups and learn from each other. In the following episode, Ms. Gatto describes various strategies she uses to develop students' understanding of multiplication concepts and their fluency with the multiplication tables.

In this episode, Ms. Gatto shows how her support of students to learn their facts is embedded in understanding multiplication concepts. She also builds on the work they have already done with addition facts. Because she uses a variety of representations, models, and contexts in her teaching, she provides multiple entry points for students who may not do well learning facts in isolation. When her students, in this instance Sandra and Lanaya, work together, Sandra uses a familiar context to help Lanaya have a way to think about number relationships that will increase her knowledge of facts.

Voices From the Field

Fact Practice

In my class, we work on fact practice throughout the year. I find that giving the students the opportunity to practice their facts when they have finished their work allows for movement as well as the opportunity to work with a partner if they choose. They can find a quiet spot in the room to work or remain at their own seats. The students have fact cards to practice throughout the year; the facts they are practicing change depending on what their needs are as well as what concepts we have covered at any given time during the year. For example, we begin the year practicing addition facts, and then move to subtraction, multiplication, and division as the year goes on. The students can practice whatever fact cards they feel they need to work on. If other students are also done with their work, they can choose to work together and test each other on the facts.

During the multiplication unit, I particularly focus on helping students become fluent with their multiplication facts. Throughout this unit, the students are working on different models to build their understanding of multiplication. Students who may not be strong at memorizing, but have visual strengths, particularly benefit from these models and representations. Representations and strategies the students work on include pictures of things in groups, skip counting, related problems, and arrays. When we work with arrays, students are encouraged to see smaller arrays within the array they are focusing on.

(Continued)

(Continued)

For example, when looking at a 6×4 array, the students might notice that this array is made from a 6×2 and another 6×2.

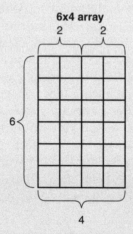

6x4 array

Other students might know $5 \times 4 = 20$ and they just need to add on another 1×4.

Here Ms. Gatto uses what she knows about her students and offers them multiple ways of learning their facts, recognizing that models and representations will help some of her learners who have strong visual spatial skills, but may struggle with memorization.

Around the middle of the unit, when they have a solid foundation, I give the children a multiplication chart and their fact cards to begin memorizing their facts. The multiplication fact cards include a space for students to write clues that will help them learn the fact (Russell et al., 2008, pp. 66–68).

We use the ideas and representations we have worked on as a starting point for our clues on our multiplication fact cards. The work we have done with the addition cards carries over to multiplication. For example, students have learned during addition practice that learning facts and using clues will help them with more challenging problems. So now with multiplication, knowing 4×6 can help with 8×6 by doubling or knowing 6×5 can help with 8×5 by adding on 2 more groups of 5 (perhaps even sketching an array on the card).

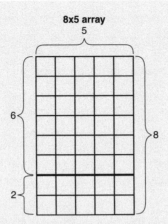

8x5 array

The clues the students generate for their cards are individualized. As I check in with the students, I am able to see what types of clues the students are using to be sure these clues will help them toward fluency with the facts.

To help the students who are struggling with the memorization of particular facts, I may ask another student to work with them a couple of times throughout the week. The students are encouraged to make two piles of their facts: "combinations I know and combinations I am working on" (Russell et al., 2008, pp. 64–66). The partner can then help with the clues for the combinations that need work. If the clues are not helping, I encourage using story contexts. Working with a partner may help with this idea. One example of this took place with Sandra and Lanaya. Lanaya struggles to memorize her facts, despite the amount of practice she does on a daily basis. I had asked Sandra to work with her one day and they came up with a story context that has helped Lanaya. Sandra and Lanaya were working with their multiplication cards and Sandra was trying to help Lanaya work through several problems, one of which was 3 × 6. I happened to sit in on the following conversation:

Sandra: It's like 3 groups of 6. So we could say that there were 3 kids and each kid had 6 pieces of candy.

Lanaya: Yeah, I think that makes sense. I could draw 3 people and give them each 6 dots for the candy.

Sandra: I don't think Ms. Gatto wants you to draw a picture all the time though. That might take a long time. Maybe we could skip count instead.

Lanaya: Okay, 6s are hard to count by, but I do know that 6 and 6 is 12. So 6 more would be. . . . [after counting on her fingers] 18.

(Continued)

(Continued)

Sandra: So 6 times 3 is 18. That would be 18 pieces of candy.

Lanaya: Oh, that makes sense. The 18 is all the candy together.

Ms. Gatto: Do you think that story will help you work through other problems you are struggling to remember Lanaya?

Lanaya: [Silent at first] I think so. It's like the first number will be the number of people and then I can skip count to stand for the candy they have. I always get confused with my groups of things. Maybe this can help.

Sandra: I think it can work. See, another card you have is 8 times 4. That would just be 8 people and 4 pieces of candy each. Then you can count the 4 pieces of candy for each person.

Lanaya: So . . . for this one 7 times 6, that's still hard. But, I think that means there are 7 people and 6 pieces of candy for each one. I can count by 6s even though that's hard. Maybe I should write that down. . . . Oh, the first one we did was 3 times 6. I might be able to use that for 7 times 6 'cuz they are both 6s.

I wanted to intervene during this conversation, but decided to let them do all the talking; aside from my one question (I couldn't resist). Sandra was able to help Lanaya see the relationship of several different problems using the same context. Although the problems will still be challenging for Lanaya, perhaps the idea of people having pieces of candy will help her visualize the equal groups. She commented that she gets confused with her groups of things, so maybe this idea will help her keep track of what the numbers mean in each of the problems she is practicing. I was also quite impressed with Lanaya's ability to see the relationship of 3 × 6 and 7 × 6. She recognized that she already figured out 3 × 6 and pointed out that both problems involve 6s, hopefully connecting that to the 6 pieces of candy for each person, 6 in each group.

In Ms. Gatto's classroom, students feel comfortable asking for help, admitting confusion, and helping each other. Students know both from watching Ms. Gatto and from her explicit modeling that "helping" someone doesn't mean telling them the answer, but rather listening to their thinking, asking questions, and offering representations or contexts that might make sense to the students. Ms. Gatto also helps students reflect on their strategies, asking Lanaya, for example, if contexts might help her in other situations.

Since this conversation, I have seen Lanaya continue with this strategy when working on her multiplication facts. Although she has not yet fully mastered the facts, she has this context to fall back on when she gets confused with the meaning of the factors and what to skip count by. She has avoided drawing pictures since Sandra pointed out the time involved in that. Her skip-counting skills are being reinforced and she is beginning to memorize more of her "combinations I am working on."

On occasion, to be sure that everyone has the opportunity to practice their facts, I will allow time for the whole class to practice. During this time, I will ask that they record the "combinations I know" and "combinations I am still working on" in their math notebooks. This activity lets the students know that they are responsible for their own independent learning, but I will be checking in with them to see how they are doing. I also then have the opportunity to see a snapshot of how each student is progressing, which skill they should revisit, what the next skill I should focus on is, or what future work might require additional practice.

Playing "math around the world" from the beginning of the year is another big part of our fact practice. During the game, two students are presented with a math fact and the first one that says the correct answer moves around the circle to the next person, and they both try to be first to answer the next fact question. Our goal is for one person to make it around the entire class without missing a fact. The students know when the next big date for our math around the world will be and it motivates many of the students.

As a class, we discuss how the students can prepare themselves for this ongoing game. Typically the students will say they need to practice their facts, at home and at school. As we get into our curriculum for the year, the students begin to understand how memorizing their facts is important, but they also need strategies to figure out facts that they haven't memorized yet. So, the students will begin to use their fact cards to assist them. When the big day arrives, the students are excited and eager to begin.

I also give my students timed tests on occasion to check in on their progress. The idea isn't to see who did the best in the class, but what improvements each individual student made from one test to the next. Typically, we keep these timed tests in a student folder so students can each monitor their own progress. Allowing the students to keep track of their own progress on these tests then allows them to go back to their fact cards, pull out the facts they didn't get on the test, and consistently practice these facts. When I administer the next timed test, students check to see which of the facts they struggled with before are now correct.

Ms. Gatto provides a variety of structures throughout the year to help her students become fluent with facts. Students understand that they are expected to practice and learn the facts, both working independently and through the activities she provides on a regular basis.

Her approach focuses on putting them in charge of their own learning, and taking responsibility for learning the facts. Her math class centers on making sense of mathematics, and all students participate in solving problems, whether or not they have learned their facts. However, the students understand that learning the facts will make it easier for them to solve math problems and work toward that goal.

Some technology resources can help teachers assess their children's mathematical knowledge and provide customized lessons for students to develop in areas that are weak. Such individualization can be helpful for teachers who are overwhelmed by the diverse needs of their students. Programs such as the National Library of Virtual Manipulatives (http://enlvm.usu.edu/ma/nav/index.jsp) and DreamBox Learning (www.dreambox.com) can help students visualize relationships and give them feedback that can help them identify mistakes.

Summary

The examples in this chapter illustrate how teachers think carefully about their students' approaches to learning, instead of thinking about their struggles to memorize facts in isolation. They analyze when and how the student is having difficulty and what seems to help them learn mathematics. They plan how to provide the necessary supports, including a variety of contexts, models, and representations. Further, they help students make connections to prior knowledge, and encourage them to take responsibility for their own learning and reflect on what helps them learn. Even at a young age, in the case of Ms. Walker's students, the exposure to mathematical facts must be situated in ideas and concepts. Ms. Gatto's students keep track of the facts they are working on, write clues to help them with facts, and work in pairs to help each other master the facts. In Darrell's case, both his parents and some of his teachers noticed his visual and spatial strengths and his ability to learn from contexts to help him overcome his obstacles with memory.

Notes

1. An array is a rectangular arrangement of quantities in rows and columns, for example, a dozen eggs, a six-pack of juice cartons.

2. Children use the doubles facts (starting with the 10 doubles facts from $0 + 0$ to $9 + 9$) to solve problems. Doubles plus one, or near doubles, includes all combinations where one addend is one more than the other. The strategy is to double the smaller number and add one.

6

Building Students' Abilities to Plan, Organize, and Self-Monitor in Mathematics Class

Ms. Thompson: I am going to send you off now to work on problems. I am going to give you two problems to choose from. Both are fifth-grade problems; one is easier and one is harder. You are going to work on your whiteboards. If you already have two solid ways of solving multiplication problems, you might want to challenge yourself with the harder problem. If you were stuck on a second way, you might want extra practice with an easier problem. You need to decide which one you want to choose.

Ms. Thompson's directions to her inclusive math class are explicit, and come with the expectations that her students will be able to self-monitor, in this case to make informed choices about what problems and strategies are appropriate for them, and have a way to get started. She did not expect her students to be able to do this in the beginning of the year. These directions, from March of the school year, come after many opportunities she gave the students to think for themselves about their own learning, to plan and organize their work, all with the supports and encouragement she provided.

Ms. Thompson's approach is different from what often happens to help students who are struggling with planning, organizing, and **self-monitoring**. Too often, students who need help with these skills are given mnemonic devices, told to use key words, but are not given experiences that allow them to think for themselves and to practice the skills and processes they need to build and integrate. If students are only taught to solve problems by identifying key words, they tend to solve problems incorrectly in which these words are misleading (Lester & Kroll, 1994). If they have been taught only to look at the numbers and not think about the operation to use, they often do not attempt to make sense of the problem (Booker, Bond, Briggs, & Davey, 1998; Kaur & Blane, 1994). The reliance on key words and procedures often leads to a dependency on adults for help, the *learned helplessness* that can become a habit for students with learning disabilities. This dependency and lack of self-confidence interferes with students' ability to take control of their own learning, to develop the skills such as setting goals, planning effective strategies for reaching the goals, self-monitoring methods and performance, managing time efficiently, taking responsibility for results, and adjusting future strategies accordingly. They do not develop a sense of personal agency or confidence that they can learn (Zimmerman, 2002).

Special educators have turned a great deal of attention to the skills involved in planning, organizing, and self-monitoring. These skills are important in students' ability to achieve. Some researchers have found that a child's ability to self-regulate in the early years predicts his or her reading and mathematics achievement better than IQ scores (Blair & Razza, 2007; Pape & Smith, 2002; Schunk & Zimmerman, 1998). There is some debate in the literature about which terms to use to describe the processes that underlie these skills: among them, executive function, metacognition, and self-regulation. Executive function is most frequently described as an umbrella term for the "complex cognitive processes that serve ongoing, goal-directed behaviors" (Meltzer, 2007, p. 2). Included in these processes are self-regulation and metacognition, as well as cognitive flexibility, working memory, and attention/inhibiting impulsive behavior.

The most commonly used definition of metacognition refers to students' knowledge about how to learn and how to use regulatory strategies to manage their own learning (Desautel, 2009; Schunk & Zimmerman, 1998). Self-regulation is defined to include motivation and affective/emotional factors such as behavioral monitoring and self-control, and a sense of personal agency. It is the self-directive process by which learners actively transform their mental processes into academic skills (Zimmerman, 2002). Some of the differences in which terms are used comes from their origin. Executive function is the term that, originally, cognitive neuropsychologists and cognitive psychologists more commonly used, whereas educational and educational psychology researchers more commonly used the term self-regulation (Harris, Reid, & Graham, 2004). Recently, executive function has become the dominant term among educators as well. Instead of distinguishing among these terms, we use the terms *plan, organize,* and *self-monitor* in this chapter because they clearly describe what students need to do every day in mathematics class.

The teachers, like Ms. Thompson, who have contributed to this chapter provide their students with opportunities to make sense of mathematics, which develops their abilities to plan, organize, and self-monitor. The examples and episodes illustrate how these teachers model and facilitate ways for students to plan, develop, and evaluate their own mathematical strategies. In these teachers' classrooms, students are asked to use mathematical reasoning, and explain and justify strategies. This kind of focus on making sense of mathematics "facilitates the necessary forethought, self-monitoring, and self-reflection that are crucial to self-regulation" (Pape & Smith, 2002, p. 99). As students gain proficiency in these skills and processes, in turn, their grasp of mathematical concepts develops. Sense-making in mathematics and building planning, organizing, and self-monitoring skills are intertwined.

Providing Entry Points

Some students have difficulty knowing where to start; they break down when asked to solve problems on their own. It is hard for them to know how to begin, what is the information they already know that can help them, and how to develop a sequence of steps that will lead to a solution. Ms. West, a first-grade teacher, uses a guided math group[1] to allow students to think through together about how to solve the type of problem the class will work on during the day's lesson. Ms. West carefully plans ahead of time for her students who need support to make sense of the problem. She knows which students might need real objects to understand what the problem asks. She gathers together a small group who might need to "act out" the problem. The other students are able to work on a problem independently for the few minutes that she works with the guided math group.

Guided Math Group:
Introducing the Problem With First Graders

Ms. West: When I was packing, I decided to pack toys so I could play at the beach. Do you like toys? Wiggle your thumb. [They all do.] I put 10 baby dolls in my bag. Then I went to the beach and met boys and girls. They played with the dolls.

I gave 5 to the little boys and girls to play with. Who remembers my story?

Gloria: You had 10 baby dolls, and kids asked if they could play.

Ms. West: What happened next?

Gloria: You gave them 5.

Ms. West: Someone else?

David: You had 10; boys and girls came and played; you gave them 5 baby dolls.

Ms. West: At the end of my story, did I have more than 10 or less than 10?

David: Less: you took 5 away and you have 5 more.

Ms. West: Who else has an idea if it's more or less?

Alberto: Because 10 is more than 5?

Ms. West: Raise your thumb if I have less than 10. [They all do.]

Corinna: Because 4 is less than 5. And 5 is more than 1, 2, 3, 4.

Ms. West: Where is the 4 coming from?

Corinna: I think 4 is left.

Ms. West: Let's figure it out.

Ms. West passes out cutout stick figures.

Ms. West: How many have to go in the bag?

Everyone says 10.

Ms. West: Then what happens?

Mark: You gave 5 away.

Ms. West: How many are left? Can you show me on your fingers?

Everyone holds up 5 fingers.

Ms. West: Did I put baby dolls in or take them out? Should our number be more or less than when we started?

Ms. West introduces the story problem using verbal prompts and questions, representations using baby dolls and counting by fingers to guide these students through the process. Her main goal is to help the children make sense of the information and give them some entry points to use as they solved the problem. Over time, she hopes that the questions she poses are ones they will ask themselves as they make sense on their own. Her interaction with the small group also helps her to see who might need even more assistance. Corinna, who at first said 4 were left, is a child she will make sure to check in with as she sends the students off to try some problems on their own.

Using Technology

Software can also provide students who are struggling with structure to help them approach a problem. The INK-12 NSF project, "Teaching and Learning Using Interactive Ink Inscriptions in K-12" created a "stamp factory" on a tablet computer to support students' thinking in multiplication problems such as, *I have 5 cats. Each cat has 4 legs. How many legs are there?* Each stamp depicts an object (in this case, a cat) with some number of identical parts (in this case, 4 legs). Sometimes the problem includes the stamp, which the student can then use to create the right number of cats—and then count the legs. A more complex problem requires the students to draw the stamp, and then use it to illustrate and solve the problem. Drawing the stamp becomes a separate subtask for the problem and can be a way for students to focus on the central relationship in the problem (i.e., Each cat has 4 legs.).

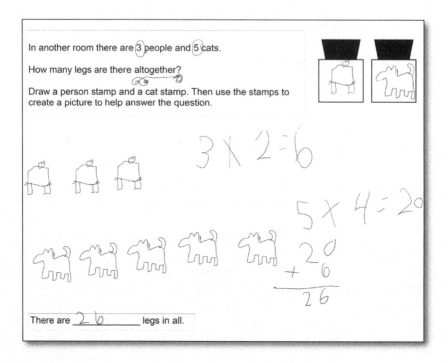

In another room there are ③ people and ⑤ cats.

How many legs are there altogether?

Draw a person stamp and a cat stamp. Then use the stamps to create a picture to help answer the question.

$3 \times 2 = 6$

$5 \times 4 = 20$

$$\begin{array}{r} 2\,0 \\ +\ 6 \\ \hline 2\,6 \end{array}$$

There are __2 6__ legs in all.

Connecting to Prior Knowledge With Third Graders

Students who need help with planning and organization also need prompting to recall what resources and tools are available to help them solve problems, and what strategies the class has previously discussed. In Ms. King's class, the students have just weighed two fruits and now their task is to find the difference between the weights (Russell et al., 2004, pp. 24–25). Ms. King reviews with her students what can help them find the difference:

Ms. King: What kind of tools do you have to help you find the difference? What's one thing that can help you?

Benjamin: The 100 chart.

Ms. King writes down what he says.

Ms. King: You have a 100 chart on the wall and one in your books. It will be easier for you to keep track of the numbers if you use the one in your books.

She shows them where the 100 chart is in their books.

Ms. King: Who can show me how to use the 100 chart to find the difference between 93 and 78? Where would you put your finger and where would you move?

Jana: I would start at 78 plus 10. That brings you to 88 and then plus 5 and that brings you to 93.

Ms. King: She didn't count by 1s; what was her first step?

Malik: She jumped up 10.

Ms. King: Then she didn't want to jump another 10 because she knew that 98 was too far. What did she do?

Jamahl: She counted up 5.

Ms. King: How else could you do it?

Jamahl: You could also go up plus 2 from 78 to 80, and then plus 10 to 90. Then plus 3 to 93.

Ms. King: So Jamahl counted up plus 2 from 78 to 80, and then plus 10 to 90 and plus 3 to 93, and Jana started out by adding plus 10 to 78 to get to 88, and then she counted 5 more to 93. Do both ways work?

The class says yes.

Ms. King: I noticed that I asked you for the difference between 78 and 93, but you counted up. Is that easier?

The class says yes.

Ms. King: What else would help you?

Ella: The 300 chart.

Ms. King: Yes, sometimes if the weight is more than 100, the 300 chart will be helpful. Sometimes we draw something, do you remember what helps us that we draw?

Ricardo: A number line?

Ms. King: Yes, we draw a line sometimes. [She writes 78 and 93 on an open number line.] How would we show the difference on the number line?

Ricardo: I would go 10 to 88 and then 5 more. That is 15 all together.

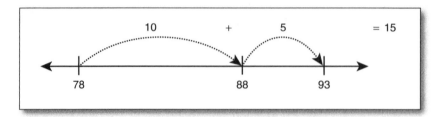

Ms. King draws the jumps Ricardo describes. She shows that they can also jump plus 2 and plus 3 instead of plus 5.

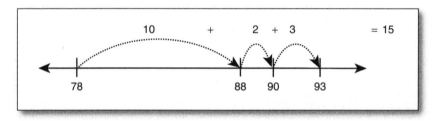

She also draws another student's method, with the jumps from 78 to 80, and 80 to 90, and 3 to 93.

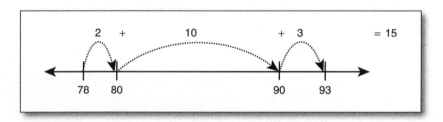

Then, she compares these methods with the strategies the class used previously on the 100 chart.

Ms. King: Could you use cubes?

The class says yes.

Ms. King: If you think you need them, you can use them. When I am collecting the scales, what are you going to be doing?

Ella: Finding the difference.

Ms. King: There are empty sheets in your notebooks for scrap paper.

In this example, Ms. King not only reviews the available tools with the students, she also reviews the strategies they might use. While this approach is useful for all of her students, her attention to explicit details is particularly targeted to her students who have difficulty knowing how to approach a learning task.

Intervention With a Second Grader

Sometimes the carefully designed group work described in the previous examples is not enough, and the student requires some one-on-one support. The following episode describes an intervention that Ms. Ingalls carries out with a student, Ramon, who has difficulty planning and organizing his work. Finding an entry point is particularly frustrating for him. She thinks carefully about what Ramon knows and where he needs support to understand the mathematics involved in the activity.

Voices From the Field

Organizing and Planning

One of the standards I have to teach my second graders each year is to determine the value of a collection of coins that includes half-dollars, dimes, nickels, pennies, and quarters. This learning goal can be difficult for many students, particularly those with issues related to planning, organizing, and self-monitoring. Counting and adding an amount of coins requires a good amount of background knowledge about the concept of money—the value of the coins, their names, and appearances. The most challenging hurdle, however, seems to be devising a strategy for organizing the counting and adding the coins and carrying out that plan. These were difficulties that I had not anticipated in my first year of teaching.

Early in my first year of teaching, I introduced collections of coins to my class. I had anticipated that keeping track of the counts might be an issue for some children, so I showed the class how to determine the value of a coin collection by drawing a number line and representing counting on that value as a jump on the number line. We practiced together by adding collections of coins that I drew on the whiteboard. As I circulated around the room, I discovered one of my students balled up in the corner, his partner trying in earnest to get him to play the game. When I asked Ramon what was wrong, he said sadly, "I don't get money."

(Continued)

(Continued)

When I first sat down with Ramon that day, I felt guilty for allowing him to get to that point of frustration, and second, realized just how complicated it truly was to add up the values of a collection of coins.

Generally, figuring out the value of a mixed collection of coins efficiently and accurately requires the counter to have a strategy ahead of time, something that seemed difficult for Ramon. The counter first has to be able to recognize the coins and know their values, and then make a plan for counting and adding them and also keep track while counting. It is often easier to keep track of the amount you have if you add the largest amounts first (such as counting all the quarters, and then adding on dimes, nickels, and finally pennies). While students often naturally come to this conclusion, students who struggle with issues related to organizing and planning may need explicit instruction in this area.

When Ramon said, "I don't get money," my first task was to figure out where his understanding of this skill was breaking down. I also had to think about all the skills that are needed to play this game. To start, I placed a collection of manipulative coins in front of Ramon and asked him to identify their names and values. He correctly named and gave the value for each coin. Next, using a 100 chart, I asked him to place each coin where it belonged on the chart, one at a time. He placed the coins correctly where they belonged on the chart (the half-dollar on the 50, the quarter on the 25, etc.). He was pleased when I pointed out to him that he already knew a lot about money. I wanted him to experience success before we practiced skills that were harder for him.

Having a sense of himself as a learner is an important step for Ramon in being able to self-monitor. Ms. Ingalls is able to show him that he already has specific knowledge.

Next, I practiced skip counting with Ramon to see if that was where his understanding was breaking down. While he was unable to skip count by 25s, he could successfully skip count by 5s, 10s, and 50s, a necessary skill when adding collections of coins. We practiced adding collections of one value of coin, such as five nickels, which he could do successfully.

Ramon's understanding of money seemed solid until he was presented with a collection of a variety of coins. I had given him one half-dollar, one quarter, one dime, one nickel, and one penny. At this point, he began to show frustration and I realized it was time to stop and rethink how best to support Ramon.

Ms. Ingalls practices skip counting with Ramon because she knows he has some strengths in this area. It also allows her to see where his understanding breaks down.

When I looked back on how Ramon had performed with the money tasks, it became clear to me that he was unable to make a plan for organizing the money to make the adding manageable. Clearly, the strategy I had taught of drawing a number line was not an effective one for Ramon. As a first-year teacher, I am still figuring out when it is helpful to offer a particular strategy and when it is better to observe how students are figuring out the problem and ask them to evaluate and compare their ways. However, in Ramon's case, I needed to figure out how to help him find a strategy that he could go to when encountering these types of problems so that whenever he saw a coin problem, he had a plan that he could execute. That plan would first need to help him identify the easiest order to count and add the coins and then a way to keep track as he counted.

In order to keep track as he totaled the amount, I thought Ramon could use a 100 chart and money manipulatives. If he took out the coins he was trying to count and actually placed them on the 100 chart as he was adding, he might be able to keep track of the amounts. It would also build his understanding of how much each coin was worth.

Next, I showed Ramon this strategy by adding a quarter, a nickel, and a penny. I placed the quarter on the 25 on the chart, then counted up five more, and put down the nickel. Finally, I counted one more space for the penny and that was my answer, 31 cents.

1	2	3	4	5	6	7	8	9	10
11	12	13	14	15	16	17	18	19	20
21	22	23	24	🪙	26	27	28	29	🪙
🪙	32	33	34	35	36	37	38	39	40
41	42	43	44	45	46	47	48	49	50
51	52	53	54	55	56	57	58	59	60
61	62	63	64	65	66	67	68	69	70
71	72	73	74	75	76	77	78	79	80
81	82	83	84	85	86	87	88	89	90
91	92	93	94	95	96	97	98	99	100

(Continued)

(Continued)

Ramon seemed delighted with this strategy. I presented him with the same collection from the day before: a half-dollar, a quarter, a dime, a nickel, and a penny. Ramon picked up the penny and put it on the one. Then he took the nickel and counted up five spaces and placed it on the chart. Ramon seemed to be doing fine until he got to the larger coins, like the half-dollar and the quarter. Adding on those amounts proved cumbersome, and Ramon often lost count and had to go back and recount. In the end, he got off track with the strategy and came up with 81 instead of 91.

Ms. Ingalls: What made you choose the penny first?

Ramon: [Shrugs] I just did . . .

Ms. Ingalls: What was the hardest part of counting the money?

Ramon: Once there got to be a lot to count it was hard.

Ms. Ingalls: So the half-dollar was the hardest?

Ramon: Yes.

By questioning Ramon about his choices, Ms. Ingalls helps him develop the ability to self-monitor, and at the same time communicate that she expects him to make sense of the task.

I asked Ramon to think about starting with the half-dollar. I pointed out that if he placed the half-dollar on the chart first, then he would not have to count up to 50 at all.

Ramon said, "Oh yeah . . ." with a smile. He put the half-dollar on the 50. I prompted him to put the remaining coins in order by their value (it is important not to say "from biggest to smallest" or student's may assume you mean the size of the coins) and count on the remaining coins. Once he was organized in this manner, he was able to successfully count on and get the correct sum. I could see his confidence building.

Because Ramon struggles with finding entry points on his own, I practiced this one strategy with him over the course of several weeks, until he would naturally take out a 100 chart and manipulative money when he encountered coin problems. He needed to be reminded to organize his coins before counting.

Later, I offered him another way to make this strategy even more effective. First I showed him how to think of the coins as 10s and 1s, for example, a quarter is two 10s and five 1s. I then showed Ramon that to count on a quarter on the 100 chart, he could just skip down two rows on the 100 chart and move over five spaces instead of counting on all 25. I made sure that he knew that each row is 10. We counted together in the beginning. This made the strategy even more efficient and accurate for Ramon.

Here, Ms. Ingalls not only gives Ramon a procedure, she solidifies his knowledge of the number system, and his fluency with 5s and 10s. By practicing what works for him at this point, she continues to help him build his sense of himself as a learner. By pointing out and naming the strategies he was using, she hopes that he will eventually be able to call upon these strategies himself.

In order to help Ramon, I had to analyze what was needed to be successful with the mathematical task, and then learn what he knew and what was hard for him in order to make a plan. It had been clear when I first sat down with Ramon that he had a strong foundation when it came to money—he could recognize coins and understood their values. I talked with Ramon about what he knew so he understood that he already had knowledge of coins. It was when he was presented with a task that needed planning and organization that he needed explicit instruction and support.

Once I decided on a plan, I introduced the strategy to him, and then practiced it with him over time. I made sure that he understood the mathematical concepts because I wanted Ramon to engage in the same mathematics as my other students. Later in the year, when I revisited counting collections of coins with my class, I asked Ramon to explain his strategy. After repeated practice he was able to so, and he was able to take ownership of his work. The class called it "the Ramon strategy," which delighted him. This strategy proved helpful to other students who were having difficulty counting money and often I would see children going to the math "tool shed" and taking out the 100 chart and manipulative money. I make a variety of resources and tools, the tool shed, readily available to my students, and I might ask them before I send them off to work independently what might help them solve the problem.

Now I teach the Ramon strategy to my whole class when I first introduce determining the value of a collection of coins, but I make sure to also refer to the mathematics of the strategy, so it becomes *the Ramon strategy: starting with the largest quantity.*

Once students have solidified this strategy, I introduce other mathematical ideas. The strategy of always counting the largest to the smallest value

(Continued)

(Continued)

of coin is not effective in all situations. For example, when counting a quarter, a dime, and a nickel, it may be easier to add the quarter and the nickel first to make a multiple of 10. I try to build this awareness in students by highlighting the 10s column on the 100 chart and setting it as a goal for students to reach that 10s column using the coins they have.

More broadly, I find it important to think about what I am teaching ahead of time and try to think like a student who struggles in math. What might be confusing? How can I give them a structure to support them before they are frustrated? When I end each focus lesson, I encourage students who are feeling "thumbs in the middle" or "thumbs down" about the concept I just presented to stay and work with me on the rug. As I work with these students, I encourage them to articulate what is confusing to them and share any strategies they devise for completing the task at hand.

Ms. Ingalls makes it acceptable to be confused or have questions as part of her classroom culture. Here, she explicitly talks about the strategies she uses to help students become aware of their own learning.

For students like Ramon, whose issues stem from the organization and planning piece, I make sure to repeatedly practice strategies that work for them until they are part of their routine. By providing him with support on the organizational and planning piece, I could see that Ramon was able to access the mathematical concepts. My hope is that with practice, Ramon will be able to develop organizational and planning strategies for himself. As he thinks of himself more as a mathematical learner, he is able to listen and learn from others' strategies. I often find that the students in my class already have some of the best strategies, and giving them time to share with one another during the focus lesson and the lesson wrap-up has become an important part of my teaching practice.

Ms. Ingalls episode reveals the intentionality of her teaching. She recognizes the skills Ramon already has and notes the areas in which he needs support. In addition to identifying his strengths, she makes sure to let Ramon know that he has strengths, relieving some of his frustration. She provides him with a sequence of lessons that take what he knows and extend his knowledge to develop a strategy for organizing his work—a strategy that he understands and that is efficient for him. Her extra support is in service of his understanding the mathematics, and she includes him in the classroom mathematics discussions. Eventually, she will use what works with him during the money unit to help him find entry points and solution strategies for himself.

Self-Monitoring: Students Evaluating Their Own Work

In addition to understanding how they best learn and what confuses them, students who struggle with self-monitoring also often need help evaluating their own work. How many times do teachers observe that the students who need support with self-monitoring do not have a good idea about how to check the reasonableness or accuracy of their solutions? Reviewing students' solutions with them is only effective if the students are actively engaged in the process and understand what they are looking for in regard to the reasoning, the organization, and the accuracy.

In the following example, a fifth-grade teacher works with a small group of students who need support in mathematics and are just beginning to work with factors and multiples. She asks students to compare their classmates' solutions in solving the problem, "What are the factors of 750?" She lists the solutions anonymously, but during the conversation, some students are comfortable identifying their own solutions.

Ms. Miller lets the class know that she and the math coordinator went over the class math assessments, and that in general they had done well on the factor pair problems. However, she wants the class to notice if all the factors were listed, how they determined the factors, and how they were organized. In preparation for the discussion, Ms. Miller thought carefully in advance about which student samples to choose.

Ms. Miller: Today, we will go over your explanations. I want you to particularly notice how each of the students organized their factor pairs. I also want to check in to see how you figured out which numbers are factors.

Student 1
Factors of 750

15×50

1×750

10×75

2×375

5×150

25×30

6×125

Ms. Miller: How did this person start?

Belinda: They took 50, which is an easy number, and figured out there are 15 50s in 750.

Ms. Miller: So 50 is an easy number; what did the student do next?

Elena: They did 1 times 750; that is easy.

Ms. Miller: The student has written 1×750. How do you think he or she figured out 10 times 75? Is this an easy one or hard one, Elena?

Elena: Easy, because all you have to do is add the 0.

Ms. Miller: So 10 times 75 is multiplying 75 10 times, and you know that is 750? How could you figure that out if you didn't know it?

Elena: You could break up 75 into 70 and 5 . . . 10 times 70 is 700 and 10 times 5 is 50 so that is 750.

Ms. Miller: That is a good way to do it. Breaking up the numbers into parts, you can multiply easily and see that 10 times 75 is a factor pair.

Manuel: And 10 is an even number. All the even numbers are easier: 15 is hard, and 50 is easy.

Ms. Miller: What about 25?

Lamar: That is easy because you can think of quarters.

Belinda: I see a pattern; they're all even times odd.

Ms. Miller: That is an interesting observation, Belinda. We'll have to see if that is true with other numbers.

Ms. Miller: Has the student tried all of the numbers? How can you tell?

Elena: They are kind of out of order.

Ms. Miller: What do you mean, Elena?

Elena: Well, Student 1 starts with 15 times 50, but then it's 1 times 750, and 10 times 75. There is no order. He left out 3 times 250.

Ms. Miller: Let's look at another one together. Let's notice how the factor pairs are organized.

Ms. Miller asks questions that focus the students on her mathematical goals.

Student 2
Factors of 750

1×750

5×150

6×125

10×75

50×15

2×375

Ms. Miller: Is 1 a good place to start?

Most say yes.

Ms. Miller: Yes, every number has 1 and itself as a factor pair.

What did this person do? How did they think of doing it? Are there some numbers left out?

Lamar: They skipped 25.

Ms. Miller: What can you multiply times 25 to get 750? Lamar, you said you use quarters to figure out how many 25s make up a number.

Lamar: Yeah, 4 quarters in 100, so there would be 28 in 700 and two more in 50 so it's 30 times 25 is 750.

Ms. Miller: That's a good way to figure it out. Did this person skip anything else? Is it easy to tell?

Rianna: It's hard for me to think from looking at it. It starts out with 1, 5, 6, 10, but then goes to 10 and 50 and 2 is at the end.

Ms. Miller: How do we make it easy to tell if we have left numbers out? Let's look at another example to see if we can figure out what helps us know if we have all of the factors.

Again, Ms. Miller structures the conversation so that students are considering their strategies and practicing their skills, for example, counting by 25s.

Student 3
Factors of 750

1×750

2×375

3×250

4 crossed out

5×150

6×125

crossed out 7, 8, 9

10×75

15×50

Lamar: He did good.

Ms. Miller: Why do you say that? Lamar, explain what you think he did. [Manuel has acknowledged that he is Student 3.]

Lamar: He listed them in an organized way—the hundreds were going down.

Ms. Miller: So the numbers on the right side go from largest to smallest. What about the numbers on the left side?

Jake: The numbers there go from smallest to largest.

Ms. Miller: So he made a list of numbers on the left hand side in sequence, trying out each one and figured out if they were factors and then crossed out some. Why are 7, 8, 9 crossed out?

Lamar: Because 7 is not a factor of 750 and neither is 8 or 9. I tried them.

Ms. Miller: How did you know to skip to 15?

Manuel: I looked at the factors of 600, 700, 800 to help me [looking at the factor pair tables Ms. Miller has posted in the room]. I saw that 12 times 50 is 600, and 14 times 50 is 700. I figured out that 13 times 50 would be 650, so I knew those numbers wouldn't work with 750.

Ms. Miller: How did you know 15 times 50? Is that easy?

She has the class count by 50s out loud because she wants to make sure that all of the students understand at least one way to figure this out.

Ms. Miller: How did you know that 50 would fit in the list of factor pairs?

Lamar: Because the number is 750.

Ms. Miller: So the numbers are in order, and Manuel also used what he knew, the factors of 600, 700, and 800 to help him.

Ms. Miller: All of you did well. What we need to work on is explanations, and we need to work on having a clear organization.

In this example, Ms. Miller uses students' own work to get them thinking about organization as a way to listen to what they understand about factors. Since most made a good start on listing the factor pairs, she is able to steer the conversation toward the students' explanations of how they approached the problem. She actively involves the students by asking them questions: *How do you know? What do you mean? Are there some numbers left out? Is 1 a good starting place?* She brings out their number knowledge: *How did you know to skip to 15? How could you figure that out if you didn't know it?* She also points out the strategies that students were using: So, the numbers are in order, and Manuel also used what he knew, the factors of 600, 700, and 800 to help him. Her next step might be to help students figure out factors through number relationships: *If 3 × 250 is 750, how could your knowledge of multiples help you figure out other pairs (for example, 6 × 125 = 750)?*

Software programs have also been developed that can be useful in helping students learn to evaluate their work. Programs, such as those produced by DreamBox Learning (www.dreambox.com), provide targeted feedback. For example, the computer can record representations of a student's strategy and prompt the student to use that strategy with subsequent problems. Students also learn when to use the Help features provided by the software, such as representations and tools (100 chart, number lines).

Gradual Release of Responsibility

The previous episodes illustrate the supports to offer students when they need help with planning, organizing, and self-monitoring. In the following episode, Ms. Gordon is able to move one of her students, Emma, along the road to organizing her work, planning strategies, implementing a solution, and monitoring her own learning. She describes her work with Emma, who, like Ramon, needs help with entry points and solving problems on her own. Ms. Gordon includes Emma along with other students in mini-lessons that focus on making sense of the problem, and also provides her with structures and a series of steps that help her plan, organize, and evaluate her work.

Voices From the Field

Becoming an Independent Problem Solver

One of the biggest challenges of teaching students who have difficulty planning and organizing their work is helping them to succeed on their own. Often, you'll work one-on-one with a student—like my third grader Emma—and everything will go well. With my attention and scaffolding, Emma succeeds and shows me she really understands the math. But then, I have to leave Emma to help other kids. Ten minutes pass and when I come back, Emma is staring at her paper and has not solved a single problem since I left.

Without me next to her, the various steps the task requires overwhelm Emma. Like many kids with these organizational issues, Emma has trouble focusing on what she needs to solve, making a plan for attacking the problem, keeping numbers straight in her head, and showing her solution. As a result, she circles around the problem, reading it over and over again, unable to get started.

How do you give enough support to students like Emma without neglecting the rest of your class? And what tools and strategies will help her do the work on her own?

I begin by providing for a gradual release of responsibility in my math workshops, which helps all learners, not just Emma. The workshop starts with a mini-lesson, such as one I taught on story problems that require subtraction. As a whole class, we built an anchor chart of "subtraction situations"—take away, comparing, and finding the missing part problems—and put them next to a chart we made earlier on "addition situations." I wanted students to be able to read story problems, and use an appropriate strategy to solve them. They often refer to the chart to help them get started.

When students begin working on their own, I identify who needs small-group support. For this lesson, I planned ahead that I would include Emma in my group because word problems require many steps and are so tricky for her. I also knew Josh and William would need help, because of the nature of their learning disabilities.

I announce that I'm having a small group, and anyone is welcome to join us. Dana and Collin got up right away because they each read the first problem and didn't know what to do. I help kids to develop this self-awareness over the course of the year so no one is left staring at the ceiling, not asking for help.

Ms. Gordon has established a classroom community in which children feel comfortable asking for help. The small-group structure she provides gives students a safe and comfortable way to get support.

With my group at the side table, I start by providing a lot of scaffolding—in this case, by reviewing the content of our mini-lesson—and then I gradually decrease the scaffolding as the students' understanding improves.

This is the time when I can help Emma establish habits of mind that will enable her to be successful on her own. In the small group, we practice problem-solving steps (on a checklist for each student): (1) Understand the problem. (2) Think of a strategy. (3) Make a plan and try it. (4) Check: does your answer make sense?

This subtraction lesson was tricky because students really had to understand a problem before they knew how to solve it. That was where this group needed extra support. Together, we read the first problem and visualized what was happening.

Emma: It says, "Mr. Jackson's class collected 212 cans for recycling. Their goal is to collect 300 cans. How many more cans do they have to collect to reach their goal?"

(Continued)

(Continued)

Ms. Gordon:	Thanks, Emma. On our checklists, the first problem-solving step is to understand the problem. One of the best ways to do this is to use visualization, just like we do when we read books. I want you all to reread the problem silently, and when you do, make the "movie in your mind" and give a thumb's up when you can visualize what's happening.
William:	I can see the class with lots of bags of cans, but they don't have enough yet.
Ms. Gordon:	Thanks, William. What else are you visualizing?
Collin:	I'm visualizing that they need a lot of cans to make 300. [Emma, Dana, and Josh are nodding and giving silent applause. It looks like we're on the same page.]
Ms. Gordon:	Thanks, Collin. So, what do we need to solve in this problem? Everyone, find the question in the problem and underline it.
Emma:	The question is, "How many more cans do they have to collect to reach their goal?"

Ms. Gordon provides her group with an explicit way to enter the problem that builds on their experiences with reading. Along with the visualizing, students are explaining what they see.

Josh:	How many more! It's a comparing problem. You need to compare how many cans they have now with how many cans they need to meet their goal.
Ms. Gordon:	Give silent applause if you're thinking the same as Josh [vigorous silent applause from everyone in the group]. So, now we can check off "Understand the problem." What's the next step on our checklist?

Using techniques like silent applause that won't interfere with the rest of the class members who are working independently, Ms. Gordon checks in on whether or not students are following along and recognizing themselves whether or not that they are understanding what the problem is asking.

When we come to Step 2, think of a strategy, most of the kids will pick from the variety of subtraction strategies we've learned as a class. But, with kids like Emma, who easily get lost in their steps, I like to reinforce a single "go-to strategy." Thinking of multiple strategies can be overwhelming, so at first we try to help them develop a single strategy that they understand and can generalize with larger numbers. Eventually, my goal is for my students to use a variety of strategies and to become flexible problem solvers.

For Step 3, make a plan and try it, a tool that helps everyone, but particularly students like Emma, is the acronym ESAL. When making a plan for their solution, my students all write ESAL on their papers. It stands for equation, strategy, answer, label: all the steps you need to show in your work. It also gives Emma a way to work through her subtraction strategy without getting lost or trying to keep too many numbers and ideas in her head.

While ESAL is a mnemonic device, it is not a substitute for students' thinking, but rather serves as a way for students to organize their approach.

While the rest of the group works on the problem, I work with Emma one-on-one.

Ms. Gordon: So, what's your go-to strategy for subtraction?

Emma: Adding up on the number line.

Ms. Gordon: Great. How can you make a plan?

Emma: ESAL! [She writes ESAL down on her paper, and then stares at the E for a long moment, then writes down 300, and then erases it.]

Ms. Gordon: Tell me about your thinking.

Ms. Gordon does not jump in with a suggestion. To build Emma's ability to self-monitor, she encourages her to express what she is thinking.

Emma: I don't know which number comes first. Do I start with 300 or 212?

Ms. Gordon: That can be confusing. When you get confused, I want you to visualize what's happening in the problem. What happened in the movie in your mind?

(Continued)

(Continued)

Emma: They had 212 cans. [She writes down 212.] And they needed to add more cans to equal 300. [She adds to her equation: 212 + _____ = 300.] Now I can add up on the number line. I got it.

Emma checks off the E in ESAL and is off and running with her strategy. If she gets lost in her steps, she can always come back to her equation to get refocused. Once she's done with the strategy, she checks off the S. She writes her answer (checking the A), and finally labels and checks off the L. Then, Emma can return to her problem-solving checklist, make sure her answer makes sense, and move on to the next problem.

When my small group is solving a problem, I leave them for a moment and circulate around my class, quickly scanning students' work. I don't stop for long, and I try to avoid the temptation to teach anyone one-on-one. Instead, I maximize my impact by checking for understanding and, if needed, sending additional students to the group for help. I also make mental notes of who might need help the next day.

Ms. Gordon keeps track of the progress, not just of her small group, but of her whole class. She has built in routines so that students expect to work independently at times, and also expect her to check on them while they're working and pose questions to monitor their understanding.

I also take these opportunities to gather students who I observe need an additional challenge to help them extend their learning. I can quickly give them directions, send them off to work, and then return to my small group.

Once a student is confident and ready, he or she leaves the group and works on his or her own while I continue to work with the other students. The boundary between kids getting extra help and kids working on their own is fluid throughout the math workshop.

Over the course of a typical math lesson, the small group gradually decreases in size until only a few students need support or everyone is working independently. After getting lots of support in the beginning of the math lesson, Emma gets less support as we work and relies more on using her problem-solving checklist and ESAL.

On her addition and subtraction end-of-unit assessment, Emma did all of her work independently, and earned a near-perfect score. She beamed when I handed back her test. When I asked her what she thought was the key to her success, she told me, "Practice! And ESAL and using the checklist. Math used to be so hard. But now, I just follow the steps."

> Emma made it sound so simple: just follow the steps! She has come a long way from the nervous math student who would spin her wheels, reading problems over and over, unable to get started. What I think her statement really shows is that Emma has always been capable of doing the math. Her strategies have given her a map to follow through solving any problem, and practice has given her the confidence she needs to take each step.

Emma experiences success in Ms. Gordon's room because Ms. Gordon builds on her strengths, and offers her structured support, while expecting her to make sense of mathematics. Ms. Gordon also structures her classroom so that it is acceptable for students to ask for help and have the small group to join if they need help. Some of the scaffolds she uses for Emma involve a checklist and a mnemonic device; however, the focus of her intervention with Emma remains on understanding the operations of addition and subtraction and what each problem is asking. Although Emma says she "just followed the steps," each step in the sequence requires her to make sense.

Summary

The approach of the teachers in this chapter contrasts with what usually happens with students who struggle with planning, organizing, and self-monitoring. Often the students are given graphic organizers, mnemonic devices that emphasize procedure, but not understanding. Not only does this reinforce the students' learned helplessness, it also is a one-size-fits-all approach. In contrast, the strategies these teachers use, whether it is giving students explicit choices like Ms. Thompson, presenting a series of student work examples to help students evaluate solutions like Ms. Miller, or providing a sequence such as ESAL that Ms. Gordon uses, these methods are always in service of understanding the mathematics. As students engage in mathematical sense-making, it builds their abilities to plan, organize, and self-monitor, and that, in turn, allows the students to be better able to solve mathematical problems independently. Research findings confirm that encouraging students to engage in self-monitoring has produced positive results. Meta-analyses of effective strategies for working with students who have difficulties in math reveal that helping students think aloud—verbalizing, drawing, and writing their solution strategies—produced a consistently large effect size of .98 (Gersten & Clarke, 2007). The verbalization appeared to anchor students and decreased the impulsivity that sometimes impedes students' ability to problem-solve. Other studies showed that focusing on helping students set goals, identify the relevant features of problems, and verify and reflect on their solutions, improved children's mathematics learning (Fuchs et al., 2003; Muir & Beswick, 2005).

The examples in this chapter are from classrooms in which teachers understand not only what mathematics students need to work on, but they also determine what students already understand. They make sure to let the students know that they already have some knowledge and encourage students to articulate what is hard for them or what confuses them. They review the available resources and strategies that the class has discussed. They pose questions to help students evaluate their work: *Is this a good way to start? Is there anything you left out? How can you figure something out if you don't know it right away?* They encourage all students to express confusion as well as agreement with a solution, to take risks, and not be afraid to say, "I don't know." While these teachers offer support and in some cases model useful organizing strategies, they also actively engage their students in thinking mathematically. By focusing on sense-making, students' strengths, and setting high but reasonable expectations, they help their students see themselves as learners of mathematics, a first step in working on the skills they need to be able to approach, solve, and evaluate their solutions to mathematical problems.

Note

1. A guided math group is a structure based on the principles of guided reading groups in which a teacher supports students' development of math skills, strategies, and concepts within the context of a small group. Teachers facilitate this learning through focused conversations and targeted questioning.

Conclusion

What does it look like to teach mathematics that focuses on sense-making to a range of students in inclusive classrooms?

The preceding chapters provide a framework and examples to address the question posed in the introduction. The examples offer teachers resources and strategies to help them meet the challenge of teaching mathematics to diverse learners, and to dispel some common myths:

1. There are "tips" and "procedures" that can make teaching in inclusive classrooms straightforward.

2. Students with special needs cannot learn mathematics unless they are told what to do.

3. "Good teaching" for all children is enough for students with special needs to succeed in mathematics.

Complexity of Teaching in Inclusive Mathematics Classrooms

Teachers who understand the complexity of teaching in inclusive classrooms have written each "Voices From the Field" segment. The process is anything but simple. As Karp and Voltz (2000) state,

Being a successful weaver of lessons for diverse groups of students requires the ability to integrate effectively what is known about the content (in this case, mathematics), what is known about how to teach it, and what is known about the students to whom it is taught. (p. 27)

A major theme of each chapter is how purposeful these teachers are—how thoroughly they plan in order to meet their students' needs. They not only plan for each lesson, but they also plan to build the *culture of acceptance* that is the foundation of an inclusive mathematics classroom. Ms. Gordon establishes a routine for extra help earlier in the year, announcing to the class, "At the round table, I'm going to do some more problems like this with a small group. If you'd like some help getting started, please come join us."

As a resource teacher, Ms. Tarlow makes it a point to interview students if she observes that they are struggling with a math concept. By spending this time with the students, she learns what they know, what they still need to practice, and how they approach a task. Ms. Tran uses sentence frames as a starting point to help her students who are learning English get accustomed to expressing their mathematical ideas as they learn mathematical terms. These sentence frames are open-ended, giving students a structure that facilitates their thinking. All of these teachers show a command of math content, pedagogical strategies, and knowledge of their students.

Making Sense of Mathematics

All of the teachers who contributed to this book expect their students to make sense of mathematics, and they provide supports and resources that facilitate their students' understanding of mathematical concepts. They might use technology, as Mr. Daniels does when he makes a fraction number line on the SMART Board that his student can use flexibly as she masters the concept of equivalent fractions. They might develop structures, as Ms. Gatto does with her regularly scheduled "turn-and-talks," during which students talk in pairs about a particular concept. Ms. Gatto takes her students through a carefully orchestrated sequence so that they are aware of the purpose of and expectations for their roles during turn-and-talk. These strategies are very different from what often happens for students with special needs in mathematics class. Too often, they are presented with a series of steps and procedures to follow and not engaged in mathematical discourse that helps build their ability to grasp mathematical ideas.

Making Mathematics Explicit

"Good teaching for all" is not enough to help students with special needs succeed in the inclusive math classroom. As stated previously, the voices from the field in this book reveal the purposefulness and intentionality these teachers bring to their practice. While they expect that their students with special needs can and will learn mathematics, they also realize that they need to put in place supports and meticulously sequenced activities and lessons to make the mathematics explicit and the expectations for class assignments clear. Ms. West knew that some of her first-grade students would struggle to solve a story problem on their own without acting it out first. She plans a guided math group and acts out the problem with real objects first, asking questions, and naming the strategies students use to indicate the important mathematics in the problem. Ms. Thompson knew that her student, Michael, who needed support with cognitive flexibility, did not apply his knowledge of place value when he used rote procedures, and became confused when trying to solve problems in multiple ways. She observes and thinks through the times that Michael has been successful. After noting that he could break numbers apart and multiply the parts, she helps Michael develop his strength into a reliable strategy. When Ms. Miller asks her students to evaluate each other's solutions, she communicates her goals and expectations clearly to the students: "I want you to particularly notice how each of the students organized their factor pairs. I also want to check in to see how you figured out which numbers are factors." Throughout the lesson, she emphasizes the mathematics and clarifies students' comments about their organization and justification for deciding which numbers are factors and multiples.

In all of these examples and the many others in the book, the emphasis is on sense-making, along with providing the resources, tools, and structures that support the students to develop mathematical understanding. The focus on making sense implies that the students themselves are doing the thinking, that is, it is not coming from adults who "spoon feed" them. When Ms. Gatto decides to interview her students about what they learned from her turn-and-talk sessions (in which partners discuss a mathematical strategy, make a prediction, or evaluate an answer to her problem), Berlyn, one of her students who struggled with attention, shows remarkable insight. He comments that talking with his classmates helps him get his ideas out, and they work out ideas together. He adds, "I think it keeps me focused." The students described in this book, like Berlyn, are taking responsibility for their learning. It is only when students see themselves as learners can we "count them in," as members of a mathematical learning community.

Glossary and Resources

Ableism

Ableism is a term that refers to society's discrimination against people with disabilities, also known as physicalism, handicapism, and disability oppression.

Useful Resources and References

Griffin, P., Peters, M. L., Smith, R. M. (2007). Ableism curriculum design. In M. Adams, L. A. Bell, & P. Griffin (Eds.), *Teaching for diversity and social justice* (2nd ed.). New York: Taylor & Francis.

Hehir, T. (2005). Eliminating ableism in education. In L. I. Katzman (Ed.), *Special education for a new century.* Cambridge, MA: Harvard Educational Publishing Group.

Abstract Thinking

The *American Heritage Stedman's Medical Dictionary* (2002) defines abstract thinking as "thinking characterized by the ability to use concepts and to make and understand generalizations, such as of the properties or pattern shared by a variety of specific items or events." In math, an example is generalizing about number patterns or properties of operations. Young children are often able to engage in abstract thinking earlier than was previously believed. However, children with certain disabilities, such as autism spectrum disorder, tend to have difficulty with abstract thinking.

Useful Resources and References

Abstract thinking. (2002). In *American Heritage Stedman's medical dictionary.* Boston: Houghton Mifflin Company.

Seo, K. H., & Ginsburg, H. P. (2004). What is developmentally appropriate in early childhood mathematics education? Lessons from new research. In D. H. Clements, J. Sarama, & A. M. DiBiase (Eds.), *Engaging young children in mathematics: Standards for early childhood mathematics education* (pp. 91–104). Hillsdale, NJ: Lawrence Erlbaum.

Steen, L. A. (1999). Twenty questions about mathematical reasoning. In L. Stiff (Ed.), *Developing mathematical reasoning in Grades K–12* (pp. 270–285). Reston, VA: NCTM Yearbook.

Accommodations

Accommodations are changes that teachers make in the way tasks are taught or the classroom environment is structured so that children with disabilities can learn along with their classmates.

Useful Resources and References

National Dissemination Center for Children with Disabilities (NICHCY) (http://nichcy.org/)

National Center for Learning Disabilities (NCLD). (2006). *Accommodations for students with LD.* Retrieved from http://www.ldonline.org/article/Accommodations_for_Students_with_LD

Asperger's Syndrome

Asperger's syndrome is a developmental disability characterized by normal intelligence, motor clumsiness, unusual and intense interests, a limited ability to appreciate social nuances and develop friendships, impaired nonverbal communication such as facial expressions and body language, and strong preference for routine and consistency. The *DSM-V* (American Psychiatric Association, 2012) is now including Asperger's under autism spectrum disorders.

Useful Resources and References

National Institute of Neurological Disorders and Stroke (www.ninds.nih .gov/disorders/asperger/detail_asperger.htm)

OASIS@MAAP (www.aspergersyndrome.org)

American Psychiatric Association. (2000). *Diagnostic and statistical manual of mental disorders (DSM IV).* Washington, DC: Author.

Attwood, T. (2005). *What is Asperger's syndrome?* Retrieved from www .aspergersyndrome.org/Articles/What-is-Asperger-Syndrome-.aspx

Attention Deficit Hyperactivity Disorder (ADHD)

Attention deficit hyperactivity disorder is a neurological-based condition that is characterized by the following behaviors that occur over a period of time: distractability, short attention span, and impulsiveness. There are three subtypes: (1) predominantly hyperactive-impulsive, (2) predominantly inattentive, and (3) combined hyperactive-impulsive and inattentive. Most children have this combined type of ADHD.

Useful Resources and References

National Institute of Mental Health, U.S. Department of Health and Human Services (www.nimh.nih.gov/health/topics/attention-deficit-hyper activity-disorder-adhd/index.shtml)
National Resource Center on ADHD (www.help4adhd.org)
U.S. National Library of Medicine (www.nlm.nih.gov/medlineplus/attention deficithyperactivitydisorder.html)
American Psychiatric Association (2000). *Diagnostic and statistical manual of mental disorders (DSM IV).* Washington, DC: Author.
Flick, G. L. (2010). *Managing ADHD in the K–8 classroom.* Thousand Oaks, CA: Corwin.

Autism Spectrum Disorder

Autism spectrum disorder is a developmental disability that usually begins in infancy or early childhood. The characteristics include deficits in social responsiveness and interpersonal relationships, abnormal speech and language development, and repetitive or stereotyped behaviors. In addition, the children with autism often have atypical sensory responses, such as to certain sounds or textures. Each of these symptoms varies from mild to severe. They will present in each individual child differently.

Useful Resources and References

National Institute of Mental Health, U.S. Department of Health and Human Services (www.nimh.nih.gov/health/publications/autism/what-are-the-autism-spectrum-disorders.shtml)
American Psychiatric Association (2000). *Diagnostic and Statistical Manual of Mental Disorders (DSM IV).* Washington, DC: Author. (See www.dsm5 .org for proposed changes in the next edition.)

Cognitive Flexibility

Flexible cognition entails the ability to adapt to a variety of task demands, to consider and respond to multiple aspects of a task, and to consider multiple representations of an object or event.

Useful Resources and References

Deak, G. O. (2003). Flexible problem solving in children. *Advances in child development and Behavior, 31,* 271–326.

Homer, B. D., & Hayward, E. O. (2008). Cognitive and representational devel-opment in children. In K. B. Cartwright (Ed.), *Literacy processes: Cognitive flexibility in learning and teaching* (pp. 19–41). New York: Guilford Press.

Spiro, R. Feltovich, P., & Coulson., R. L. (2004). Cognitive flexibility theory. In *Theory into practice database*. Retrieved from http://tip.psychology.org/spiro.html

Common Core State Standards Initiative

As detailed on the website, www.corestandards.org, the Common Core State Standards Initiative is a state-led effort coordinated by the National Governors Association Center for Best Practices (NGA Center) and the Council of Chief State School Officers (CCSSO). Teachers, school administrators, and experts collaborated on the development of the standards in English language arts and mathematics. "The K–5 stan-dards in mathematics were designed to provide students with a *solid foundation in whole numbers, addition, subtraction, multiplication, division, fractions and decimals*—which help young students build the foundation to successfully apply more demanding math concepts and procedures, and move into applications" (para. 1).

Useful Resources and References

Common Core State Standards Initiative (www.corestandards.org/about-the-standards/key-points-in-mathematics)

Hunt Institute (www.hunt-institute.org)

National Council of Teachers of Mathematics (www.nctm.org/standards/mathcommoncore/)

Council for Exceptional Children (CEC). *Common Core Standards: What special educators need to know.* Retrieved from http://www.cec.sped.org/AM/Template.cfm?Section=CEC_Today1&TEMPLATE=/CM/ContentDisplay.cfm&CONTENTID=15269

Differentiation

Tomlinson (2001) notes,

> Differentiation consists of the efforts of teachers to respond to variance among learners in the classroom. Whenever a teacher reaches out to an individual or small group to vary his or her teaching in order to create the best learning experience possi-ble, that teacher is differentiating instruction. (para. 2)

When planning for differentiation, teachers need to consider the classroom: environment, curriculum, assessment, instruction, and classroom management.

Useful Resources and References

K–8 Access Center (www.k8accesscenter.org/index.php/category/math/)
Hall, T., Strongman, N., & Meyer, A. (2011). *Differentiated instruction and implication for UDL.* Retrieved from http://aim.cast.org/learn/history archive/backgroundpapers/differentiated_instruction_udl.
Tomlinson, C. A. (2001). Differentiation of instruction in the elementary grades. In *ERIC Digest.* Retrieved from http://www.ericdigests .org/2001-2/elementary.html

Disproportionality

Disproportionality is the "inappropriate overidentification or disproportionate representation by race and ethnicity of children as children with disabilities" (U.S. Department of Education, n.d., para. 2).

Useful Resources and References

The Civil Rights Project (http://civilrightsproject.ucla.edu)
U.S. Department of Education, Office of Special Education Programs (n.d.). *Topic: Disproportionality.* Retrieved from http://idea.ed.gov/explore/ view/p/,root,dynamic,TopicalBrief,7,

Executive Function

The National Center for Learning Disabilities (2010) describes executive function as "a set of mental processes that helps connect past experience with present action. People use it to perform activities such as planning, organizing, strategizing, paying attention to and remembering details, and managing time and space" (para. 1).

Useful Resources and References

Meltzer, L. (Ed.). (2007). *Executive function in education: From theory to practice.* New York: Guilford Press.
Council for Exceptional Children. (2008). *Improving Executive Function Skills—An Innovative Strategy that May Enhance Learning for All Children.* Retrieved from http://www.cec.sped.org/AM/Template.cfm?Section=Home&CONTENT ID=14463&CAT=none&TEMPLATE=/CM/ContentDisplay.cfm
National Center for Learning Disabilities. (2010). *What is executive function?* Retrieved from http://www.ncld.org/ld-basics/ld-aamp-executive -functioning/basic-ef-facts/what-is-executive-function

Expressive Language Disability

"Children with an expressive language disorder have problems using language to express what they are thinking or need" (Medline Plus, 2010, sec. 2, para. 3). These children may struggle with organizing their thoughts, constructing sentences, finding the right words when speaking, have limited vocabulary, and leave words out. Children who are learning a second language may struggle to express themselves because they are in the process of language acquisition. They are sometimes misdiagnosed as having an expressive language disability.

Useful Resources and References

American Speech-Language-Hearing Association (www.asha.org)

National Clearinghouse for English Language Acquisition (www.ncela.gwu
.edu/development)

Medline Plus. (2010). *Language disorder—children.* Retrieved from http://
www.nlm.nih.gov/medlineplus/ency/article/001545.htm

Gradual Release of Responsibility

The gradual release of responsibility model represents a gradual transition from teacher's modeling of problem-solving strategies, for example, to the student's responsibility for demonstrating and articulating the use of a particular strategy.

Useful Resources and References

Teaching Literacy in the Turning Points School (www.turningpts.org/pdf/
Literacy.pdf)

Fisher, D., & Frey, N. (2008). *Better learning through structured teaching: A framework for the gradual release of responsibility.* Alexandria, VA: Association of
Supervision and Curriculum Development.

Keene, E. O., & Zimmermann, S. (1997). *Mosaic of thought: Teaching comprehension in a reader's workshop.* Portsmouth, NH: Heinemann.

Individualized Education Program (IEP)

The Individualized Education Program (IEP) is the centerpiece of the Individuals with Disabilities Education Act (IDEA). It is a process that is intended to ensure educational opportunity for students with disabilities. The IEP is an agreement that guides and documents specially designed instruction for each student with a disability based on the student's unique academic, social, and behavioral needs.

Useful Resources and References

U.S. Department of Education (http://idea.ed.gov/explore/home)

Christie, C. A. & Yell, M. L. (2010). Individualized education programs: Legal
requirements and research findings. *Exceptionality, 18*(3), 109–123.

Council for Exceptional Children. (2005). *Council for Exceptional Children's initial
summary of selected provisions from Part B Proposed Regulations for the
Individuals With Disabilities Education Act.* Arlington, VA: Author.

Council for Exceptional Children. (2012). *Federal outlook for exceptional children: Fiscal year 2012.* Arlington, VA: Author.

U.S. Department of Education, Office of Special Education and Rehabilitative
Services, *Thirty-five Years of Progress in Educating Children With Disabilities
Through IDEA,* Washington, D.C., 2010.

Interactive Whiteboard

An interactive whiteboard (e.g., SMART Board) consists of an LCD projector that projects a computer desktop onto an interactive board. The easiest way to envision the interactive whiteboard is that the board becomes the computer and it is controlled with the teacher's hands or an interactive pen that can write and highlight documents and presentations on the computer. It allows for teachers and students to interact with the software on their computers. Most interactive whiteboards supply supplemental software that enhances the board with many other options including rulers, protractors, charts, graphs, and so forth.

Useful Resources and References

National Clearinghouse for Educational Facilities (www.ncef.org/rl/interactive_whiteboards.cfm)

Least Restrictive Environment (LRE)

A least restrictive environment means that a student who has a disability should have the opportunity to be educated with nondisabled peers in an inclusive setting, to the greatest extent appropriate. To determine what setting is appropriate for a student, a school team reviews the student's needs and interests.

Useful Resources and References

National Dissemination Center for Children With Disabilities (http://nichcy.org/schoolage/placement/lre-resources)
Wrightslaw (www.wrightslaw.com/info/lre.index.htm)

Memory Difficulties

Students with memory problems in mathematics tend to have problems with either short-term memory, the active process of storing and retaining information for a limited period of time, or working memory, the ability to hold on to pieces of information until the pieces blend into a full thought or concept (e.g., reading each word until the end of a sentence or paragraph and then understanding the full content). The impairment might be more predominant with auditory memory or visual memory or both. Deficits in long-term memory (information that has been stored and that is available over a long period of time) are less likely to come up with mathematics learning for young children.

Useful Resources and References

Learning Disabilities Association of America (www.ldanatl.org/aboutld/
parents/ld_basics/types.asp)
Mercer, C. D., & Miller, S. P. (1997). Educational aspects of learning disabili-
ties. *Journal of Learning Disabilities, 30* (1), 46–56.
Sliva, J. (2004). *Teaching inclusive mathematics to special learners, K–6.* Thousand
Oaks, CA: Corwin.

Metacognition

Metacognition refers to an individual's ability to have insight into
one's own thinking and knowing, to monitor one's own learning, and
make adjustments when necessary. These metacognitive processes
help an individual make meaning, and select and revise cognitive
tasks, goals, and strategies.

Useful Resources and References

Math VIDS: Video Instructional Development Source (http://fcit.usf.edu/
mathvids/understanding/understanding.html)
Fisher, R. (1998). Thinking about thinking: Developing metacognition in
children. *Early Child Development and Care, 141,* 1–15.
Flavell, J. (1979). Metacognition and cognitive monitoring: A new area of
cognitive-developmental enquiry. *American Psychologist, 34,* 906–911.
Hennessy, S. (1993). Situated cognition and cognitive apprenticeship:
Implications for classroom learning. *Studies in Science Education, 22,* 1–41.

Nonverbal Learning Disorder (NVLD)

Nonverbal learning disorder (also called nonverbal learning
disability and NVLD) is a particular type of learning disability.
Individuals with this disability are highly verbal, with their areas of
deficit being in the nonverbal domains. The four major categories of
dysfunction that often present themselves include a combination of
learning, academic, social, and emotional issues. These may include
the following:

1. Motoric (both fine and gross motor)

2. Visual-spatial-organizational (involving recall and perception)

3. Social (difficulties in understanding nonverbal communication
   and social cues, and difficulties adjusting to new situations and
   transitions)

4. Abstract reasoning (difficulties with problem solving and
   understanding spatial concepts)

Useful Resources and References

National Center for Learning Disabilities (www.ncld.org)
NLD on the Web (www.nldontheweb.org)

Pervasive Developmental Disorder (PDD)

Pervasive developmental disorder refers to a group of disorders, also sometimes referred to as autism spectrum disorders, characterized by delays in the development of socialization and communication skills, often accompanied by difficulty with changes in routine or familiar surroundings, and repetitive body movements or behavior patterns. Autism is the most characteristic and best studied PDD. Other types of PDD include Asperger's syndrome, childhood disintegrative disorder, and Rett syndrome.

Useful Resources and References

National Institute of Neurological Disorders and Stroke (www.ninds.nih
.gov/disorders/pdd/pdd.htm)
Matson, J., & Sturmey, P. (Eds.). (2011). *International Handbook of Autism and Pervasive Developmental Disorders*. New York: Springer.

Post-Traumatic Stress Disorder (PTSD)

"Post-traumatic stress disorder is a type of anxiety disorder. It can occur after one has seen or experienced a traumatic event that involved the threat of injury or death" (U.S. National Library of Medicine, n.d., para. 1).

Useful Resources and References

National Center for PTSD (www.ptsd.va.gov/public/pages/ptsd-children
-adolescents.asp)
National Institute of Mental Health (www.nimh.nih.gov/health/topics/
post-traumatic-stress-disorder-ptsd/index.shtml)
Jaycox, L. (2004). *CBITS: Cognitive behavioral intervention for trauma in schools*. Frederick, CO: Sopris West Educational Services.
Rice, K. F., & Groves, B. (2005). *Hope and healing: A caregiver's guide to helping young children affected by trauma*. Washington, DC: Zero to Three Press.
U.S. National Library of Medicine. (n.d.). Post-traumatic stress disorder. In *PubMed Health*. Retrieved from http://www.ncbi.nlm.nih.gov/pubmed health/PMH0001923/

Response to Intervention (RTI)

Response to intervention is an initiative that integrates assessment and intervention within a multitiered prevention system to maximize student achievement and to reduce behavioral problems. Schools use data to identify students at risk for poor learning outcomes, monitor student progress, provide evidence-based interventions, and adjust the type and level of the interventions depending on a student's responsiveness, and to identify students with learning disabilities or other disabilities.

Useful Resources and References

National Center on Response to Intervention (www.rti4success.org)

Fuchs, D., Fuchs, L. S., & Stecker, P. M. (2010) The "blurring" of special education in a new continuum of general education placements and services. *Exceptional Children, 76*(3), 301–325.

Searle, M. (2010). *What every school leader needs to know about RTI.* Alexandria, VA: ASCD.

Self-Monitoring

Self-monitoring is the ability to observe and keep track of one's own academic and social behavior.

Useful Resources and References

Hallahan, D. P., & Kaurmman, J. M. (2000). *Exceptional learners: Introduction to special education* (8th ed.). Boston: Allyn & Bacon.

Rutherford, R. B., Quinn, M. M. & Mathur, S. R. (1996) *Effective strategies for teaching appropriate behaviors to children with emotional/behavior disorders.* Reston, VA: Council for Children With Behavioral Disorders.

Vaughn, S. Bos, C. S., & Schumm, J. S. (2000). *Teaching exceptional, diverse, and at-risk students in the general education classroom* (2nd ed.). Boston: Allyn & Bacon.

Self-Regulation

Self-regulation is "an active, constructive process whereby learners set goals for their learning and then attempt to monitor, regulate, and control their cognition, motivation, and behavior, guided and constrained by their goals and the contextual features in the environment" (Pintrich, 2000, p. 453).

Useful Resources and References

Collins, G. (2008). Self-regulation as related to the transfer of learning from one setting to another. *Discoveries (National Institute for Learning Development), 25*(1), 12–13.

Pintrich, P. R. (2000). The role of goal orientation in self-regulated learning. In M. Boekaerts, P. R. Pintrich, & M. Zeidner (Eds.), *Handbook of self-regulation* (pp. 451–502). San Diego: Academic Press.

Schunk, D. H. (2005) Commentary on self-regulation in school contexts. *Learning and Instruction, 15,* 173–177.

Zimmerman, B. J. (2008). Investigating self-regulation and motivation: Historical background, methodological developments, and future prospects. *American Educational Research Journal, 45*(1), 166–183.

Zimmerman, B. J., & Campillo, M. (2003). Motivating self-regulated problem solvers. In J. E. Davidson & R. J. Sternberg (Eds.), *The nature of problem solving* (pp. 233–262). New York: Cambridge University Press.

Universal Design for Learning (UDL)

Universal design for learning is a set of principles for curriculum development and teaching that give all individuals equal opportunities to learn. UDL provides a plan for developing instructional goals, methods, materials, and assessments that work for everyone—not a single, one-size-fits-all solution, but rather, flexible approaches that can be individualized. The UDL guidelines are organized according to the three main principles of UDL that address (1) representation, (2) expression, and (3) engagement.

Useful Resources and References

Center for Applied Special Technology (CAST) (www.cast.org/udl/index.html)

National Center for Universal Design for Learning (www.udlcenter.org)

Rose, D. H., & Meyer, A. (Eds.). (2006). *A practical reader in universal design for learning.* Cambridge, MA: Harvard Education Press.

References

The Access Center. (2006). *Using mnemonic instruction to teach math.* Retrieved from http://www.k8accesscenter.org

American Psychiatric Association (2000). *Diagnostic and statistical manual of mental disorders (DSM IV).* Washington DC: Author.

American Psychiatric Association (2012). *299.80 Asperger's disorder.* Retrieved from http://www.dsm5.org/ProposedRevisions/Pages/proposed revision.aspx?rid=97#

Angold, A., Erkanli, A., Egger, H. L., & Costello, E. J. (2000). Stimulant treatment for children: A community perspective. *Journal of the American Academy of Child and Adolescent Psychiatry, 39,* 975–984.

Artiles, A. J. (2003). Special education's changing identity: Paradoxes and dilemmas in view of culture and space. *Harvard Educational Review, 73*(2), 164–202.

Artiles, A. J., & Klingner, J. K. (2006). Forging a knowledge base on English language learners with special needs: Theoretical, population, and technical issues. *Teachers College Record, 108*(11), 2187–2194.

Ball, D. L., & Bass, H. (2003). Making mathematics reasonable in school. In G. Martin (Ed.), *Research compendium for the principles and standards for school mathematics* (pp. 27–44). Reston, VA: National Council of Teachers of Mathematics.

Barkley, R. A. (2002). Major life activity and health outcomes associated with attention-deficit/hyperactivity disorder. *Journal of Clinical Psychiatry, 63*(l12), 12–15.

Barkley, R. A. (2006). *Attention deficit hyperactivity disorder: A handbook for diagnosis and treatment* (3rd ed.). New York: Guilford Press.

Blair, C. & Razza, R. P. (2007). Relating effortful control, executive function, and false belief understanding to emerging math and literacy ability in kindergarten. *Child Development, 78*(2), 647–663.

Booker, G., Bond, D., Briggs, J. & Davey, G. (1998). *Teaching primary mathematics* (2nd ed.). Melbourne: Longman.

Bresser, R., Melanese, K., & Sphar, C. (2009). *Supporting English language learners in math class.* Sausalito, CA: Math Solutions Publications.

Burkhardt, S. A. (2007). Non-verbal learning disabilities. In S. Burkhardt, F. E. Obiakor, & A. F. Rotatori (Eds.), *Current perspectives on learning disabilities* (pp. 21–34). Oxford, UK: Elsevier.

Centers for Disease Control and Prevention (CDC). (2005). *What is attention-deficit/hyperactivity disorder (ADHD)?* Retrieved from http://www.cdc.gov/ncbddd/adhd/what.htm

Chapin, S. H., O'Connor, C., & Anderson, N. C. (2009). *Classroom discussions: Using math talk to help students learn.* Sausalito, CA: Math Solutions Publications.

Clements, D. H., & Sarama, J. (2010). *SRA real math building blocks preK.* New York: McGraw-Hill Education.

Currie, J., & Stabile, M. (2006). *Child mental health and human capital accumulation: The case of ADHD.* Cambridge, MA: National Bureau of Economic Research. Retrieved from http://www.nber.org/papers/w10435

Dendy, C. A. A. (2000). *Teaching teens with ADD and ADHD.* Bethesda, MD: Woodbine House.

Desautel, D. (2009). Becoming a thinking thinker: Metacognition, self-reflection, and classroom practice. *Teachers College Record, 11*(8), 1997–2020. Retrieved from http//www.tcrecord.org. ID Number: 15504.

DuPaul, G. J. (2007). School-based interventions for students with attention deficit hyperactivity disorder: Current status and future directions. *School Psychology Review, 36*(2), 183–194.

DuPaul, G. J. & Stoner, G. (2003). *ADHD in the schools.* New York: Guilford Press.

DuPaul, G. J. & White, G. P. (2004). An ADHD primer. *Principal Leadership Magazine, 5*(2), 11–15.

Dyson, A. H. & Smitherman, G. (2009). The right (write) start: African American language and the discourse of sounding right. *Teachers College Record, 111*(4), 257–274.

Fosnot, C. T. (Ed.). (2010). *Models of intervention in mathematics: Reweaving the tapestry.* Reston, VA: National Council of Teachers of Mathematics.

Freed, D. (2011). Persistent questions about attention deficit hyperactivity disorder. In A. Burstyzn (Ed.), *Childhood psychological disorders: Current controversies* (pp. 53–70). Santa Barbara, CA: Greenwood Publishing Group.

Fuchs, L., Fuchs, D., Prentice, K., Burch, M., Hamlett, C., Owen, R., et al. (2003). Enhancing third-grade students' mathematical problem solving with self-regulated learning strategies. *Journal of Educational Psychology, 95,* 306–315.

Garnett, K. (1998). *Math learning disabilities.* Retrieved from http://www.ldonline.org/article/5896

Gersten, R., & Clarke, B. S. (2007). Effective strategies for teaching students with difficulties in mathematics. *NCTM Research Brief.* Reston, VA: National Council of Teachers of Mathematics. Retrieved from http://www.nctm.org/news/content.aspx?id=8452

Hankes, J. (1996). An alternative to basic skills remediation. *Teaching Children Mathematics, 2*(8), 452–458.

Harris, K. R., Reid, R. R., & Graham, S. (2004). Self-regulation among students with LD and ADHD. In B. Y. L. Wong (Ed.), *Learning about learning disabilities* (pp. 167–195). San Diego: Elsevier Academic Press.

Harry, B., & Klingner, J. (2006). *Why are so many minority students in special education?* New York: Teachers College Press.

Hehir, T. (2002). Eliminating ableism in education. *Harvard Educational Review, 72*(1), 1–33.

Hehir, T. (2007). Improving instruction for students with learning needs: Confronting ableism. *Education Leadership, 64*(5), 8–14.

Jensen, P. S., & Cooper, J. R. (2002). *Attention deficit hyperactivity disorder: State of the science, best practices.* Kingston, NJ: Civic Research Institute.

Jensen, P. S., Kettle, L., Roper, M., Sloan, M., Dulcan, M., Hoven, C., & Bauermeister, J. (1999). Are stimulant treatment of ADHD in four

U.S. communities overprescribed? *Journal of the American Academy of Child and Adolescent Psychiatry, 38,* 797–804.

Karp, K.S., & Voltz, D. L. (2000). Weaving mathematical instructional strategies into inclusive settings. *Intervention in School and Clinic, 35*(4), 206–215.

Kaur, B., & Blane, D. (1994). *Probing children's strategies in mathematical prboelm solving.* Paper presented at the AARE Conference, University of Newcastle.

Lappan, G., Fey, J. T., Fitzgerald, W. M., Friel, S. N., & Phillips, E. D. (2004). *Connected mathematics program.* Upper Saddle River, NJ: Pearson-Prentice Hall.

LD Online. (n.d.) *Glossary.* Retrieved from www.ldonline.org/glossary.

Lesaux, N. K. (2006). Building consensus: Future directions for research on English language learners at risk for learning difficulties. *Teachers College Record, 108*(11), 2406–2438.

Lester, F., & Kroll, D. (1994). Assessing student growth in mathematical problem solving. In G. Kulm (Ed.), *Assessing higher order thinking in mathematics* (pp. 53–70). Washington, DC: American Association for the Advancement of Science.

Lipsky, D. K., & Garner, A. (1999). Inclusive education: A requirement of a democratic society. In H. Daniels & P. Garner (Eds.), *World yearbook of education 1999: Inclusive education* (pp. 12–23). London: Kogan.

Loe, I. M. & Feldman, H. M. (2007). Academic and educational outcomes of children with ADHD. *Journal of Pediatric Psychology, 32*(6), 643–654.

López, A., Correa-Chávez, M., Rogoff, B., & Gutiérrez, K. (2010). Attention to instruction directed to another by U.S. Mexican-heritage children of varying cultural backgrounds. *Developmental Psychology, 46*(3), 593–601.

Losen, D. J., & Orfield, G. (Eds.). (2002). *Racial inequity in special education.* Cambridge, MA: Harvard Education Press.

Mandell, D. S., Davis, J. K., Bevans, K. B., & Guevara, J. P. (2008) Disparities in special education placement among children with attention deficit/hyper activity disorder. *Journal of Emotional and Behavioral Disorders, 16*(1), 42–51.

McKinley, L. A., & Stormont, M. A. (2008). The school supports checklist: Identifying support needs and barriers for children with ADHD. *Teaching Exceptional Children, 41*(2), 14–19.

Mejía-Arauz, R., Rogoff, B., Dexter, A., & Najafi, B. (2007). Cultural variation in children's social organization. *Child Development, 78*(3), 1001–1014.

Mejía-Arauz, R., Rogoff, B., & Paradise, R. (2005). Cultural variation in children's observation during a demonstration. *International Journal of Behavioral Development, 29,* 282–291.

Meltzer, L. (Ed.). (2007). *Executive function in education: From theory to practice.* New York: Guilford Press.

Muir, T. & Beswick, M. (2005). *Where did I go wrong? Students' success at various stages of the problem-solving process.* Paper presented at the MERGA 2005 Conference: Building Connections: Research, Theory and Practice, Melbourne.

National Center on Response to Intervention. (2010). *The essential components of RTI.* Retrieved from http://www.rti4success.org

Neuropsychonline. (n.d.). *Cognitive rehabilitation therapy.* Retrieved from https://www.neuropsychonline.com/ncrt.pdf

Pape, S., & Smith, C. (2002). Self-regulating mathematics skills. *Theory Into Practice, 41*(2), 93–101.

Pennington, B. F. (2009). *Diagnosing learning disorders: A neuropsychological framework* (2nd ed.). New York: Guilford Press.

Rabiner, D. L., Marray, D. W., Rosen, L., Hardy, H., Skinner, M., & Underwood, M. (2010). Instability in teacher ratings of children's inattentive symptoms: Implications for the assessment of ADHD. *Journal of Developmental and Behavioral Pediatrics, 31*(3), 175–180.

Reid, R. C., & Lienemann, T. O. (2006) *Strategy instruction for students with learning disabilities.* New York: Guilford Press.

Reid, R., Casat, C. D., Norton, J., Anastopoulos, A. D, & Temple, E. P. (2001). Using behavior rating scales for ADHD across ethnic groups: The IOWA Conners. *Journal of Emotional and Behavioral Disorder, 9*(4), 210–218.

Rose, D. H., & Meyer, A. (2009). *A practical reader in universal design for learning.* Cambridge: Harvard University Press.

Russell, S. J., Economopoulos, K., Wittenberg, L., et al. (2008). *Investigations in number, data, and space* (2nd ed.). Glenview, IL: Pearson Scott Foresman.

Russell, S. J., Tierney, C., Mokros, J., & Economopoulos, K. (2004). *Investigations in number, data, and space.* Glenview, IL: Pearson Scott Foresman.

Schunk, D. H., & Zimmerman, B. J. (Eds.). (1998) *Self-regulated learning: From teaching to self-reflective practice.* New York: Guilford Press.

Scruggs, T. E., & Mastropieri, M. A. (2000). The effectiveness of mnemonic instruction for students with learning and behavior problems: An update and research synthesis. *Journal of Behavioral Education, 10,* 163–174.

Shnoes, C., Reid, R., Wagner, M., & Marder, C. (2006). ADHD among students receiving special services: A national survey. *Exceptional Children, 72*(4), 483–496.

Skiba, R. J., Poloni-Staudinger, L., Gallini, S., Simmons, A. B., & Feggins-Azziz, R. (2006). Disparate access: The disproportionality of African American students with disabilities across educational environments. *Exceptional Children, 72,* 411–424.

Stigler, J. W., & Hiebert, J. (2009). *The teaching gap: Best ideas from the world's teachers for improving education in the classroom.* New York: Simon & Schuster.

Suh, J., Johnston, C., & Douds, J. (2008). Enhancing mathematical learning in a technology-rich environment. *Teaching Children Mathematics, 15*(4), 235–241.

Swanson, H. L., Cooney, J. B., & McNamara, J. K. (2004). Learning disabilities and memory. In B. Y. L. Wong (Ed.), *Learning about learning disabilities* (pp. 41–80). San Diego: Elsevier Academic Press.

Valenzuela, J., Copeland, S., Qi, C., & Park, M. (2006). Examining educational equity: Revisiting the disproportionate representation of minority students in special education. *Exceptional Children, 72*(4), 425–441.

Waitoller, F. R., Artiles, A. J., & Cheney, D. A. (2009). The miner's canary: A review of overrepresentation research and explanations. *Journal of Special Education, 20*(10), 1–21.

Wilson, S. M., & Peterson, P. (2006). *Theories of learning and teaching: What do they mean for educators.* Washington, DC: National Education Association.

Zentall, S. (2005). Theory- and evidence-based strategies for children with attentional problems. *Psychology in the Schools, 42*(8), 821–836.

Zentall, S. (2006). *ADHD and education: Foundations, characteristics, methods, and collaboration.* New York: Merrill.

Zentall, S., Smith, Y. N., Lee, Y. B., & Wieczorek, C. (1994). Mathematical outcomes of attention-deficit hyperactivity disorder. *Journal of Learning Disabilities, 27*(8), 510–519.

Zimmerman, B. J. (2002). Becoming a self-regulated learner: An overview. *Theory Into Practice, 41* (2), 64–70.

Index

NOTE: Figures are identified by (fig.)